International Perspectives on Geographical Education

Series editors

This series is under the editorial supervision of the International Geography Union's Commission on Geographical Education. Led by the priorities and criteria set out in the Commission's Declaration on Geography Education Research, the series plays an important role in making geography education research accessible to the global community. Publications within the series are drawn from meetings, conferences and symposiums supported by the Commission. Individual book editors are selected for special editions that correspond to the Commission's ongoing programme of work and from suitable submissions to the series editors. In this way, research published represents immediate developments within the international geography education community. The series seeks to support the development of early career researchers in publishing high quality, high impact research accounts.

More information about this series at http://www.springer.com/series/15101

Clare Brooks · Graham Butt
Mary Fargher
Editors

The Power of Geographical Thinking

Springer

Editors
Clare Brooks
UCL Institute of Education
London
UK

Mary Fargher
UCL Institute of Education
London
UK

Graham Butt
Faculty of Humanities and Social Science
School of Education
Oxford Brookes University
Oxford
UK

ISSN 2367-2773 ISSN 2367-2781 (electronic)
International Perspectives on Geographical Education
ISBN 978-3-319-49985-7 ISBN 978-3-319-49986-4 (eBook)
DOI 10.1007/978-3-319-49986-4

Library of Congress Control Number: 2016960311

Printed on acid-free paper

This Springer imprint is published by Springer Nature
The registered company is Springer International Publishing AG
The registered company address is: Gewerbestrasse 11, 6330 Cham, Switzerland

Contents

List of Figures

List of Tables

Chapter 1
Introduction: Why Is It Timely to (Re) Consider What Makes Geographical Thinking Powerful?

Clare Brooks, Graham Butt and Mary Fargher

> Knowledge is power. Information is liberating. Education is the premise of progress, in every society, in every family. *Kofi Annan.*

At the start of a book which is confidently titled '*The Power of Geographical Thinking*' it is perhaps appropriate that the editors expand upon why they believe that geographical thinking can be described in this way. We also assert that linked with this explanation should be some clear statements about why we consider it timely for the question of whether geographical thinking is powerful to be asked again now.

A series of points, many of which have validity across national boundaries, can be raised to locate and position the key themes both of this chapter and of the book. These take into consideration the current nature of education policy and practice both nationally and internationally, the form and function of geography as a disciplinary subject, the recent debates about powerful knowledge and whether geographical knowledge might also be considered to be powerful, and what it means to 'think geographically'. Each of these points is explored, to a greater or lesser extent, in the text that follows.

Let us rehearse why the study of geography is still important in state schools: a question which can best be answered if one knows something about the ways in which education policy and practice has shifted in different countries over the past decade (see Butt and Lambert 2014). In political and education forums (in England) during this period there have been concerns about whether the 'most appropriate forms of primary and secondary education are *really* those that favour the teaching

C. Brooks (✉) · M. Fargher
UCL Institute of Education, London, UK
e-mail: c.brooks@ucl.ac.uk

M. Fargher
e-mail: m.fargher@ucl.ac.uk

G. Butt
Oxford Brookes University, Oxford, UK
e-mail: gbutt@brookes.ac.uk

C. Brooks et al. (eds.), *The Power of Geographical Thinking*,
International Perspectives on Geographical Education,
DOI 10.1007/978-3-319-49986-4_1

of literacy, numeracy and generic 'thinking skills" (Butt 2011, p. 4 emphasis in the original). Politically, again with particular reference to England, there has been something of a rediscovery of the importance of subject knowledge in the provision of state school education. With the election of a Conservative-led coalition in 2010, and the return of a fully Conservative administration in 2015, the focus on subject teaching has sharpened. This was soon highlighted by the new Conservative government through the publication of their White Paper 'The Importance of Teaching' (2010), which moved the education agenda away from the previous Labour government's concerns about the processes of teaching and learning. The associated development of skills and competences by students was downplayed in favour of the gaining of 'essential knowledge' through a curriculum framed by traditional subjects. Here we might legitimately consider Hirschian concepts of core knowledge, Young's insistence that schools should be 'bringing knowledge back in' (Young 2008), and Kress' (2000) belief that institutional education and curricula should no longer seek cultural reproduction, but equip young people with the necessary knowledge and skills to cope with their increasingly unstable futures. The international authorship of this volume will seek to identify the extent to which such a 'knowledge turn' is also apparent in the state education systems in their own countries.

The argument for curricula to embrace the 'targeted and meaningful' contributions of subject disciplines (ibid., p. 4) is now strong, with geography being able to make claims for developing powerful thinking through its consideration and conceptualisation of contemporary and future global issues.[1] Many geography educationists' intentions are to educate our young people for active and responsible global citizenship—helping to mould future citizens who are empowered to understand and approach (geographical) challenges with confidence, whether these are local or global. The futures dimension is important, representing an element of the powerful ways in which an enlightened geography education can assist the development of our young people.

The nature of the learners' experience of geography will to a large degree reflect whether or not the subject demands, inspires or indeed ignores the application of powerful thought. The traditional rote learning of 'capes and bays', we would argue, did little to create powerful thinking through geography—whereas the more futures' oriented, decision making approaches which focus on real life issues often does. Questions of what is educationally worthwhile—from within the parent discipline, as reflected in the subject curriculum and delivered through different modes

[1]It is easy to compile a list of issues that young people will face, and for which geography offers powerful ways of thinking, including: population growth, globalization, information movement, climate change, shifting patterns of employment, migration, sustainability, interdependence, etc. As Roberts (2011) asserts 'it is precisely because of its (the future's) unpredictability that it is worth thinking about' (p. 245)—here issues that have a geographical expression can be judged, acted upon and mediated. Nonetheless with the super-complexity of some problems we must acknowledge that geographers may need to work in an inter-disciplinary way to address them successfully.

of teaching—come to the fore. It is here that work needs to be done among a range of education stakeholders—not only parents, students, and teachers but also employers, politicians and the general public—to clarify the message and promise of geography education at school level.

Our final point is so significant that it warrants a separate section.

1.1 The Nature of Geographical Thought

For young people geography education is well positioned to speak directly about many current and future issues, and can provide powerful ways of thinking about them. As Roberts (2011) reminds us, our youngsters:

> are entitled to have access to the ways in which geography sees the world, ways which will help them make sense of both their own personal geographies and also information and ideas about the world and its people that they will encounter in the future (p. 248).

This, in part, makes a case for the continued inclusion of geography in the school curriculum in the twenty-first century. Enabling students to 'think geographically' is important, especially if we can identify and justify how this thinking is powerful in helping them to see the world in different ways—particularly because 'the powerful big ideas of geography can transform the way young people see the world' (p. 249). Lambert and Jones (2013) make a similar point when they observe that the current fashion for teachers to be critically reflective about what they teach (and we see nothing wrong with this, in moderation) may ultimately be limiting; to the detriment of teachers prioritising their essential role as subject specialists. Indeed, we may need to more strongly position teachers as public intellectuals and knowledge workers who firmly embrace 'a belief that young people can be inspired, dig deep and be 'initiated' into intellectual enquiry and the world of ideas' (p. 5). This is at the heart of geography's role in facilitating powerful thought and ideas, directed by teachers who are confident in their subject expertise.

Expecting schools to be at the forefront of the development of subject knowledge is unrealistic. The *raison d'etre* of schools and universities is different, although their aims should be complementary (see Butt and Collins 2013; Lambert and Morgan 2010; Hill and Jones 2010). It should be recognised that universities, with their key role of advancing disciplinary research, are at the core of defining and re-defining the powerful nature of the content of geography, whereas schools—with their dynamic, busy, day to day focus on teaching and learning—are less central to cutting edge subject development. However, we must acknowledge that schools have important functions in ensuring that the powerful thought that underpins geographical knowledge endeavours are recognised, passed on to, and then further explored, with young learners.

Morgan (2013) is helpful in directing our attention towards what is meant by 'thinking geographically', highlighting recent debate on this theme in the academy and noting the importance of geography educators locating such discussions within a distinctly educational frame of reference. With consideration of contributions by Gregory (1994), Bonnett (2008), Harvey (1984), Mackinder (1890), Massey (1991), Matthews and Herbert (2004, 2008), Jackson (2006) and the GA's 'manifesto' (GA 2009), Morgan refers to how geography is conceptualised as a subject and what it means to be educated 'as a geographer'. Significantly he focuses our attention on the nature of geographical thought, implicitly commenting on why thinking geographically may be powerful. Rejecting objectivist notions of knowledge, Morgan notes that geographical knowledge is socially constructed, reflecting the subjective interests of geographers. It is but a short step to considerations of Young's (2008) notions of powerful knowledge and of the strength of academic disciplines, which consist of knowledge developed, accumulated and stored as a result of the work of generations of researchers. Such hard won knowledge is powerful, it can be argued, precisely because it takes young learners beyond what they would normally experience in their 'everyday' worlds. In turn, through the consideration of the endeavours of key geographical thinkers, Morgan, drawing on Mackinder's early work, identifies that geographers recognise that they must engage in visualisation—'powerful because it … makes informed judgements, intelligent guesses, and (to) situates itself in human culture' (p. 275). Thinking geographically can also be powerful because it provides an integrating bridge between the human and natural sciences (Matthews and Herbert 2008), finds and imposes order on a seemingly chaotic world (Bonnett 2008), and deals in connections, belonging and morality—tying the geographical threads that connect people and places (Jackson 2006). Importantly, and counter to Young's (2008) contentions, both Jackson (2006) and Roberts (2013) see value in starting from the learners' everyday experiences, initially rejecting high theory in favour of empirical enquiry into cultural life.

1.2 Structuring a Response

In structuring its response to the question of why is it timely to (re) consider what makes geographical thinking powerful, this book is presented in three sections each of which is united by a set of organising concepts and principles. In this introductory chapter we do not refer to any individual author's contribution by name, rather we seek to provide a sense of the overall form of the book and the ways in which its content moves from conceptual and theoretical considerations, to more practical applications. Each section is fore-fronted with a brief overview provided by one of the editors, which makes reference to each of the chapters in that section. While it is not the intention of the editors, or authors, to deliberately provoke or confront, we are aware that some of the contributions challenge existing orthodoxies and may raise more questions about the powerful nature of geographical thinking than they answer.

We have selected the various contributions to ensure coherence and coverage of appropriate themes and issues relating to powerful thinking. The authorship is international, with a pleasing combination of relatively new writers and researchers (some of whom base their writing on recent doctoral work) alongside contributions from established researchers who either provide new material on areas of their existing expertise, or bring a fresh consideration of new problems. As such, this collection will certainly not provide a single, unified view of the nature of powerful geographical thinking—rather it will offer a range of perspectives and thoughts, which the editors will help to direct the reader towards in their comments at the start of each of the sections. It is certainly not envisaged that the text is read from 'cover to cover'—although we would be delighted if one wishes to approach it in this way. Rather that the different sections might be taken discretely for the arguments and perspectives developed within them, with connections being made between ideas and observations in each of the sections. The inter relationships between the different sections, and the contributions therein, are in part highlighted by the editors.

The contributors to this book all have one thing in common: they originally presented at least the essence of the ideas contained in their chapter to an international conference of the International Geographical Union-Commission on Geographical Education (IGU-CGE) held at the University College London, Institute of Education on 13–15 April 2015. This conference was in part organised by the editors, affording them the opportunity to see the development of the authors' ideas from their initial abstracts, through their presentations, to the development of the final chapters published here. As such, this edited collection is assuredly *not* a simple reproduction of a set of proceedings from the conference—authors were given guidelines for their writing, their work has been selected on the basis of its quality and its coherence to the themes of the book, drafts of chapters have been discussed and if necessary re written, and final editorial decisions made on the most appropriate submissions. Readers will make their own minds up about how successful we have been in this endeavour, suffice it to say that our intention as editors was always to go beyond a simple retelling of a set of conference presentations.

We have attempted to draw together a publication that is not solely concerned with the immediate, politically driven, school-centred worries about (say) pupil performance against externally set targets, or fluctuations in pass rates for high stakes examinations. This is not to deny that such concerns are real, but rather to acknowledge that there are other, equally legitimate concerns held by many involved in geography education—and particularly from within the subject discipline itself—that should also be researched, debated and disseminated. For us, these concerns go to the very heart of the age-old curriculum question of 'what to teach?' This has a particular resonance concerning whether what is taught is powerful in terms of the thinking it generates and the geographical knowledge it expects young learners to acquire. The content of this book therefore connects strongly to consideration of the discipline of geography, its expression through geography education in schools, and the very nature of powerful knowledge and thought. As such, everyday concerns about 'what works' in classrooms is not necessarily at the top of our agenda for discussion. By prioritising research into the power of geographical thinking we believe that we can also speak to contemporary debates in other subjects, and across educational frontiers, to consider issues such as the promotion of social mobility and social justice, the relationship between academic disciplines, the nature of educational aims and objectives, and about what is taught in schools. This not only opens up discussion of the philosophical basis of teaching and learning, but also about how the development of school subjects in the twenty-first century may play out.

The Power of Geographical Thinking is the first in a series entitled *Perspectives on Geographical Education* for the IGU-CGE published by Springer which has the aim of pursuing contemporary aspects of research in geography education. More specifically, the series attempts to identify aspects of research in this field that are generally considered to be important by both researchers and practitioners, but which have either remained largely unexplored or which now require more contemporary thought and action. Given the nature of the current research environment in geography education—poorly funded, under theorised, and reliant on a research base mostly consisting of academics involved in initial teacher education in higher education (a base which is currently under sustained attack from government)—it might be thought that at this time such a book is peripheral to more urgent concerns. We would argue otherwise: it is notable that at national and international conferences where geography educators meet there is a strong sense that researching the powerful nature of geographical thought and education is a pressing task.

The expected audience for this book is intentionally wide: it encompasses trainee geography teachers and experienced classroom-based practitioners, masters and doctoral students who have an interest in the powerful nature of geographical thought, and researchers in the academy and beyond. We anticipate that it will appeal both to geography educators and to academic geographers who have expressed an interest in how their subject is currently conceptualised. The book is written for those whose work sits at the interface between disciplinary knowledge, thought and education. We are acutely aware of the common pressures and constraints on geography education, and research in geography education, at both the national and international scales (see Butt and Lambert 2014). It is partly as a fillip to such pressures that this book has been written.

1.3 Conclusions

Lambert and Jones (2013), in the context of the infrastructure necessary to support teachers' efforts in curriculum development, refer to 'what appears tantamount to a collapse of confidence in subject identity of secondary school teachers of geography' (p. 6). We believe that this points to the need for further deliberation about the power of geographical thinking—for if teachers have lost confidence in their ability to draw from their parent discipline of geography in their curriculum making and day to day teaching, it is likely that the subject they convey will lack power in the eyes of the learner. Over the years others have made similar observations, with varying degrees of emphasis and concern (Marsden 1997; Standish 2009, 2012; Roberts 2011).

All subjects make claims about their relevance. Geography is not alone in such an endeavour but, as Roberts (2011) reminds us, can advance a particularly strong, indeed powerful, case for relevance as 'it is an integral part of our everyday lives' (p. 246). If we accept that the discipline of geography, and the process of geography education, have important roles to play in the preparation of young people for their future lives (and we acknowledge that this case has to be regularly advanced)—then we are duty bound to show how such knowledge and thinking is powerful. Geography, like many other subjects, has a potentially vast and rapidly changing content and so its curriculum must be both selective and flexible—this reflects observations that geography's ambitions as a subject are 'absurdly vast' (Bonnett 2008). But beyond the almost constant content selection issue we would argue that to be powerful the geography curriculum should also be contemporary and regularly refreshed. As Roberts (2011) comments 'the world is changing so what seems important and relevant now might seem less so in the future… it is (more) important to give students access to ways of thinking geographically, to enable them to see the world in different ways. The powerful big ideas of geography can transform the way young people see the world' (p. 249).

There are therefore *educational* considerations for any academic discipline; considerations that take us beyond the subject itself and which directly relate to the form that subject-based education takes. What subject teachers chose, or are expected, to prioritise is important: whether the demand is for rote learning, rather than enquiry; for skills, rather than content or understanding; for strong assessed performance, rather than the development of thinking; for tasks, rather than debates. Essentially, are the priorities educational or narrowly instrumental—and therefore do they offer opportunities for young learners to develop powerful ways of thinking?

There will always be urgent concerns that take the attention of education policy makers away from the subject disciplines; concerns that appear substantial and serious, and that are considered politically expedient to address. We do not seek to belittle the importance of students being able to read and write, to confidently participate and achieve; to have strong moral and ethical ways of being, to work together harmoniously, and to achieve happiness and advancement through their education. The study of geography speaks to much of this agenda, and is therefore powerful. However, it may be argued that geography educators—by the very nature of their subject base—work on an even larger canvas, one that does not always need to respond to the immediate, but which has a futures orientation that takes in questions about our very survival as a species (see Morgan 2011).

References

Bonnett, A. (2008). *What is geography?*. London: Sage.

Butt, G. (2011). Introduction. In G. Butt (Ed.), *Geography, education and the future* (pp. 1–11). London: Continuum.

Butt, G., & Collins, G. (2013). Can geography cross 'the divide'? In D. Lambert & M. Jones (Eds.), *Debates in geography education* (pp. 291–301). London: Routledge.
Butt, G., & Lambert, D. (2014). International perspectives on the future of geographical education: An analysis of national curricula and standards. *International Research in Geographical and Environmental Education, 23*(1), 1–12.
DfE. (2010). *The importance of teaching*. London: The Stationery Office.
Geographical Association. (2009). *Geography: A different view*. Sheffield: Geographical Association.
Gregory, D. (1994). *Geographical imaginations*. Oxford: Blackwell.
Harvey, D. (1984). On the history and present condition of geography: an historical materialist manifesto. *The Professional Geographer, 36*(1), 1–11.
Hill, J., & Jones, M. (2010). 'Joined-up geography': Connecting school level and university level geographies. *Geography, 95*(1), 22–32.
Jackson, P. (2006). Thinking geographically. *Geography, 91*, 199–204.
Kress, G. (2000). A curriculum for the future. *Cambridge Journal of Education, 13*(1), 133–145.
Lambert, D., & Jones, M. (2013). Introduction: Geography education, questions and choices. In D. Lambert & M. Jones (Eds.), *Debates in geography education* (pp. 1–14). London: Routledge.
Lambert, D., & Morgan, J. (2010). *Teaching geography 11–18: A conceptual approach*. Maidenhead: Open University Press.
Mackinder, H. (1890). On the necessity of thorough teaching in general geography as a preliminary to the teaching of commercial geography. *Journal of the Manchester Geographical Society, 6*(4).
Massey, D. (1991). A global sense of place? Marxism Today, June 24–29.
Marsden, W. (1997). On taking the geography out of geography education; some historical pointers. *Geography, 82*(3), 241–252.
Matthews, J., & Herbert, D. (Eds.). (2004). *Unifying Geography: Common heritage, shared future*. London: Routledge.
Matthews, J., & Herbert, D. (2008). *Geography: A very short introduction*. Cambridge: Cambridge University Press.
Morgan, J. (2011). *Teaching secondary geography as if the planet matters*. Oxford: Fulton.
Morgan, J. (2013). What do we mean by thinking *geographically*? In D. Lambert & M. Jones (Eds.), *Debates in geography education* (pp. 273–281). London: Routledge.
Roberts, M. (2011). Conclusion. In G. Butt (Ed.), *Geography, education and the future* (pp. 244–252). London: Continuum.
Roberts, M. (2013). Powerful knowledge and geographical education. *The Curriculum Journal, 25* (2), 187–209.
Standish, A. (2009). *Global perspectives in the geography curriculum: Reviewing the moral case for geography*. London: Routledge.
Standish, A. (2012). *The false promise of global learning; why education needs boundaries*. London: Continuum.
Young, M. (2008). *Bringing knowledge back in: From social constructivism to social realism in the sociology of education*. Abingdon: Routledge.

Part I
Theorising Geographical Thinking

The first part of this book seeks to explore ideas concerning powerful thinking and powerful knowledge. In a sense it provides the conceptual foundation for the text, for although in the following two parts (II and III) certain authors choose to elucidate the theoretical and conceptual ideas concerning powerful knowledge and thinking, the first three chapters attempt to address these more directly. In a book that includes a wide range of contributions from geography educators from around the world—each of whom provides their own 'take' on how geographical thinking is developed, often with reference to their particular national jurisdiction—the first part provides something of a theoretical anchor to the text. It is perhaps unsurprising that each of the contributors to Part I refer to the work of the sociologist in education, Michael Young (among others), whose recent influence on debates in geography education about the role of powerful knowledge has been considerable. Although this book is explicitly about what makes geographical *thinking* powerful, it is but a short and necessary step to consider the nature of powerful knowledge.

The editors would argue that the original contribution, and key strength, of this book is the interplay it achieves between theory and practice—in Part I a conceptual framework is presented, which the arguments and ideas in the second and third parts often take up in practical ways. By doing so, this introductory part provides a basis for making comparisons across and between chapters, introducing themes which are subsequently taken up elsewhere. Some authors in Parts II and III attempt to make these theory–practice links explicit for the reader.

Part I consists of three chapters, written by Graham Butt, Alaric Maude and Anke Uhlenwinkel. Each author writes from a different national perspective (England, Australia and Germany, respectively) and although their national educational settings have obviously influenced their ideas, there is a sense that their views have a resonance for those in other jurisdictions. Of the three, Uhlenwinkel writes passionately about geography education in her specific national context of Germany—but even her chapter, although grounded in the educational practices of that particular country, transcends the national setting to consider more 'universal' concepts of education, geographical disciplinary knowledge and their facility to make us think powerfully. Importantly, all authors acknowledge that they cannot, and do not, answer all the questions that can possibly relate to the nature of thinking powerfully in geography in their short chapters.

Graham Butt's chapter, *Debating the place of knowledge within geography education: reinstatement, reclamation or recovery?*, starts beyond the parameters of the discipline of geography by considering the work of sociologists in education who have debated the place, role and function of knowledge for over 50 years. He uses their intellectual work as a foundation from which to consider the ways in which the geography education community has recently chosen to engage with the 'knowledge turn' evident in the English state school system. By doing so, Butt engages with the social constructivist and social realist positions alternately adopted by Young, and others, mindful of the major contribution that sociologists in education have had in this field. It is recognised that the geography education community has recently embraced many of the ideas and concepts that sociologists have pursued—although as one would expect of any academic community, geographers and geography educationists have not blithely accepted notions of powerful knowledge. They have challenged themselves to debate what makes geographical thinking powerful (if indeed it is) and to consider whether the concept of powerful knowledge has a particular purchase for geographers. Whether powerful knowledge, either overtly expressed or merely hinted at, is identifiable and has value for curricular subjects is a tension for many whose work is represented in this book. As such, a concern for many geography educationists has been whether powerful thinking and powerful knowledge are helpful conceptualisations for curriculum makers, teachers and students. The response from educational sociologists, or at least from Young and Muller, are the concepts of Future 1, 2 and 3 curricula which Butt chooses to explore, supported by Hammond's recent work. The result is a clarification of the theoretical conceptions of powerful knowledge, the introduction of different models of curricula futures, and a consideration of their effects on geography education.

Alaric Maude, in a chapter titled: *Applying the concept of powerful knowledge to school geography*, engages with many of the ideas that are initially raised in Butt's chapter—but from a very different angle, and with the additional benefit to the reader of modelling a typology of powerful thinking/knowledge that geography educationists may wish to apply. In seeking to identify the geographical knowledge and thinking that could be considered powerful, Maude takes a bold step towards defining such concepts for practitioners in schools. He attempts to explain why some geographical knowledge can be considered powerful, identifying what form this knowledge may take, as well as the activities that can promote it. In doing so he provides conceptual clarification for geography teachers, educators and researchers by illustrating each of five types of powerful thought—while also reviewing criticisms of the concept of powerful knowledge from within the geography education community. Here definitions of powerful knowledge are seen to rely not only on their particular characteristics and what they can do, but also on their provenance from within disciplinary communities. Maude quickly acknowledges the subjective nature of what might be considered 'powerful', and identifies the legitimate conjecture about what is deemed powerful and why. His five types of powerful knowledge in geography are worthy of further consideration and debate, embracing as they do: new ways of thinking about the world; ways of analysing, explaining and understanding; knowledge that gives students power (control?) over their own geographical knowledge; the ability to follow and participate in debates; and knowledge of the world. Similar to Butt and Uhlenwinkel, he highlights tensions between academic and disciplinary knowledge, particularly with reference to school subjects, and is careful not to prescribe specific geographical content to powerful thinking. Instead, Maude prefers to indicate how students' intellectual powers can be enhanced through studying geography.

Anke Uhlenwinkel, concludes the first part of the book with a contribution titled: *Geographical Thinking: Is it a limitation or powerful thinking*? Uhlenwinkel is helpfully provocative, challenging her colleagues among the geography education community in Germany—and by extension, internationally—about their apparent denial of the power of their subject, favouring the use of geography education simply as a vehicle for promoting values education, or even care. She champions the importance of disciplinary knowledge in the geography classroom, illustrating, like Young, how such knowledge and thinking can be powerful in enabling students to understand their worlds beyond 'the everyday'. This takes students across the boundaries of the subject of geography *per se*, to explore political theory and contested concepts such as that of justice. Uhlenwinkel's arguments are compelling, with a resonance beyond the German context, as she argues (with Young and Muller) for taking students away from their everyday knowledge to knowledge which is specialized, disciplinary and differentiated from their normal experiences. Her criticism of contemporary geography education in Germany—where the subject is seen as an inter-disciplinary, integrating, cooperative handmaiden for other, more important, subjects—is not uncommon elsewhere (see Butt and Lambert 2014). Highlighting that under these conditions geographical thinking lacks power and simply pursues 'good causes', she explores the 'dangers of denying (geography's) disciplinary boundaries', investigating in turn the 'anti intellectual' (Rhode Juchtern), geo ethics (Haubrich), and natural sciences (Lethmate) influences on geography education. Uhlenwinkel concludes that at its most extreme geography education in German schools may run counter to education for democratic citizenship. By rejecting rational discourse such an education denies the prospect of demanding powerful thinking from students.

Chapter 2
Debating the Place of Knowledge Within Geography Education: Reinstatement, Reclamation or Recovery?

Graham Butt

2.1 Introduction

Firth (2011), in considering the implications for geography education of recent debates about knowledge and curriculum, starts his analysis by citing the work of Barnett (2009). Barnett asserts the necessity for educationists to debate whether knowledge should be 'reinstated, reclaimed or recovered' in the subject-led curriculum. On the face of it Barnett's suggestion appears to be rather odd—for if education is about anything it is surely about the attainment of knowledge, something primarily achieved by students facing the challenges of engaging with subject disciplines. Although Barnett mainly refers to students' acquisition of knowledge in the context of higher education, the principles of gaining knowledge apply equally strongly to education in schools.

Debates about the place of knowledge in geography education have recently become more animated—encouraged by consideration of the work of Michael Young and Johan Muller (among others), both of whom have helped to provide the intellectual stimulus for geography educationists to (re) consider the importance of knowledge in the geography curriculum (Muller 2000, 2009; Muller and Young 2008; Young 2008a, b; Young and Muller 2007, 2010). Roberts (2011) and Lambert (2011) have arguably foregrounded concerns in the geography education community about the retreat from subject knowledge, both in schools and initial teacher education, highlighting how the conceptualisation of teachers merely as skilled technicians has led to impoverished thinking about the role of knowledge in education. The educational contribution of the traditional subject disciplines has been widely debated, not least by sociologists of education who identify the consequences of teaching and learning on students' social mobility and equality of opportunity. Discussion about conceptions of knowledge take us back to the

G. Butt (✉)
Oxford Brookes University, Oxford, UK
e-mail: gbutt@brookes.ac.uk

C. Brooks et al. (eds.), *The Power of Geographical Thinking*,
International Perspectives on Geographical Education,
DOI 10.1007/978-3-319-49986-4_2

philosophers of ancient Greece, who distinguished between 'pure' (theoretical, conceptual, scientific, context independent) and 'applied' (crafts, skills, 'everyday', context-dependent) knowledge forms. From here arises the distinction between the intrinsic and extrinsic value of knowledge; the difference between 'knowing that' and 'knowing how'. We must not lose sight of such divisions. Additionally, alongside considerations of substantive subject knowledge (which are those mainly pursued in this chapter), there is also a necessity to consider forms of syntactic subject knowledge that give young people epistemic access to geography.

My intention in this chapter is to start by exploring recent debates about knowledge among sociologists in education, then to narrow the focus to explore the response from within the geography education community. I begin by looking at the development of ideas about both the place and function of knowledge towards the end of the last century.

2.2 The New Sociology of Education

Michael Young's work on knowledge goes back to the early 1970s. In *Knowledge and Control* (Young 1971) he recognises the social origins of knowledge in a book that became central to the 'new sociology of education'. Significantly this work included essays by Bernstein and Bourdieu, whose contributions on social justice and education meshed closely with the thrust of ideas being promoted by the 'new sociology'. Here the process of cultural transmission of knowledge and control was highlighted—facilitating an analysis of the transfer of both power and status afforded by particular forms of education and curricula which lay at the heart of education in modern societies (Firth 2011). *Knowledge and Control* essentially explores the relationship between knowledge, curriculum and power—offering evidence that the post war educational project in English state schools had largely failed. The curriculum that most state school students studied was based on elite, or middle class, values and views which (it was argued) proved inaccessible to working class children and ensured the maintenance of a distinct social and cultural elite (see also Willis 1977). Deeply concerned with the promotion of social justice through education, the supporters of the new sociology of education questioned the dominance of traditional subjects in state school curricula believing that they promoted neo liberal and cultural restorationist values (Rawling 2001).

However, over the course of the last half century, Young developed rather different ideas. The educational arguments he helped to advance through the new sociology of education movement, based as they were on social constructivist assumptions, he now considers to be unsatisfactory attempts to establish a sociological foundation for debate about the curriculum (Firth 2011). As Young stated recently: 'It took me a long time to recognize that freedom from the existing curriculum without access to knowledge leads nowhere' (Young 2014, p. 13). Nonetheless, despite its now recognisable flaws, the new sociology movement did succeed in challenging the mostly uncritical acceptance in England of the ideas of liberal education and the 'passing on' of time-honoured traditions of thinking (see Hirst 1972; Hirst and Peters 1970; Oakeshott 1972), opening up altered perceptions of the connections between curriculum, knowledge and power (Firth 2011).

2.3 Social Realism

So how has Young's thinking changed? In '*Bringing Knowledge Back In: from social constructivism to social realism in the sociology of education*' (Young 2008a) the importance of the contribution of knowledge to the curriculum is seen as paramount. Young argues strongly that the place, role and function of knowledge in education has been largely neglected by politicians, schools and educationists to the detriment of disadvantaged students. Indeed, the secondary school curriculum is characterised as having shifted unhelpfully towards emphasizing the preparation of young people for employment and good citizenship—something curricula have always struggled to achieve—rather than prioritising the gaining of knowledge (see Hartley 2008; Firth 2011). This downplaying of knowledge, at the time Young was writing, was clearly evident in New Labour's project to strengthen the skills-based curriculum in schools, with a concomitant impact on the status of disciplinary knowledge. The focus on 'learning more than teaching', alongside the promotion of 'National Strategies', was part and parcel of other reforms that increasingly positioned schools, and their leadership teams, as businesses that would be subject to increasingly stringent school inspections and improvement measures. Mitchell and Lambert (2015) helpfully reflect on the impact of neo-liberalism on education—which forced teachers to be more accountable (measured against externally set standards and competencies) and consequently shifted our notions of teacher professionalism. Educational policies that would supposedly help to prepare young people to work and compete in a volatile global economy have introduced:

> Notions of flexibility and soft, transferable skills (which) supported a view of the subject knowledge of 'traditional' academic subjects as outdated and of questionable 'relevance' to learners. Knowledge, in this view, was equated with *information* which could be readily accessed outside school. The school's role, rather than providing access to subject knowledge, was to facilitate learning (Mitchell and Lambert 2015).

Pondering on how the (Geography) National Curriculum has changed since its inception in 1991, Morgan (2014a) refers to a 'retreat from knowledge' and the maintenance by politicians of a 'curriculum of the dead'. In essence, Morgan's analysis of the direction of travel of the school curriculum over the last quarter century reveals that 'the 'what' of curriculum (has) seemed less important than the 'how' of learning' (Morgan 2014a). Here knowledge is viewed as a social construction, where subjects are seen as arbitrary collections of content whose boundaries are not that important—in essence, if the curriculum can be shaped to interest and motivate disengaged children into greater participation in learning, but *without* a significant contribution from subjects, then so be it. Other geography educators share Morgan's concern. Firth, commenting on the place of subjects in schools, observed that:

> (education reform) has involved narrowing the aims of education to economic and social purposes and led to a focus on competencies and skills, and the shift in learning towards personalisation and learning outcomes. In all of this, abstract, formal or disciplinary knowledge is being increasingly marginalised in the curriculum in all sectors of education and in many countries (Firth 2011, p. 143).

Mitchell and Lambert (2015) also warn against geography lessons that over prioritize 'social issues' and 'opinion forming' at the expense of gaining geographical knowledge:

> If geographical knowledge development is not a core concern of the teacher/curriculum maker the question could be asked: What knowledge is being side-lined by elevating topicality and presumed 'relevance' in the classroom? (Mitchell and Lambert 2015).

But let us not lose sight of Young's contribution. The title of Young's 2008 book is revealing—it states that the author had moved from a position of supporting the notions of 'social constructivism to social realism'. Social constructivism is generally understood to be a theory of knowledge that applies the philosophical principles of constructivism in social settings. Here groups *construct* knowledge, collaboratively creating a culture of shared understandings and meanings. It contains the proposition that knowledge, including so-called 'scientific knowledge', is neither neutral nor independent of its cultural norms and values—but is actually *socially* constructed in support of particular values and understandings; hence the link between knowledge and power. Immersion within different forms of knowledge allows one to function both intellectually and socially—the implications of elite groups defining knowledge/curriculum which may act to reproduce their own inherent advantages are therefore considerable. The objectivity, or truth, of knowledge is seen as being dependent on two dimensions: the *social* dimension, that is, the ability of the knowledge claims to gain support both within and beyond a community of experts/scholars/academics, and the *realist* dimension, that is, the coherence and validity of the ways in which knowledge can explain phenomena.[1]

Social realism describes an increasingly influential school of thought in the study of knowledge and education, which places knowledge at the core. Providing students with access to disciplinary knowledge in schools is regarded as essential—it is an issue of social justice, because people need such knowledge to conduct debates, to address problems and to inform decisions within the societies in which they live. Social realists also believe that theoretical knowledge is being marginalised in the curricula of all sectors of education—particularly through competency-based training, which still provides the dominant curriculum model for vocational education in many countries. An important question for social realists is therefore: What should we teach in our schools (and in higher education

[1]It is possible that this representation of social realism underplays ontological realism and promotes epistemological realism. The former recognises that knowledge is about something other than itself; or, put another way, that reality exists independently, beyond the discourses that help us to shape our understanding of the world.

institutions)? We can extend and refocus this question by asking: 'Is disciplinary and theoretical knowledge still important in schools?'

In principle, social realists support the production of knowledge-based curricula which promote social justice and social mobility. Here knowledge is seen as an 'object' rather than as a 'process'; which leads to the rejection of conceptions of knowledge as 'malleable' and 'arbitrary' (Morgan 2014a). Thus, by adopting a social realist stance, Young chooses to emphasize the conditions and collective practices of knowledge generation that enable communities of experts to construct knowledge. This leads to a number of beliefs about knowledge: that knowledge, truth and objectivity can be recognised as fundamentally social categories—where knowledge is a rational consensus of the best evidence, and the most powerful theories, conceived by experts; that knowledge has 'testable' explanations, which are open to challenge; and that knowledge is best organised into domains with boundaries, these being associated with specialist subject communities (which are often discipline-based). These assertions have significant implications for the positioning of knowledge within the geography curriculum, and indeed within the curricula of all school subjects.

Young's championing of the importance of knowledge in the school curriculum does not distract from his underlying concerns about the connections between knowledge and power. The anxieties expressed about the relationship between education and social mobility in the early 1970s by the 'new sociologists' are still apparent, but with a clearer articulation and distinction between what Young refers to as the 'knowledge of the powerful' and 'powerful knowledge'. The former relates to what Young once termed 'high-status' knowledge, a concept that reverberates with Bourdieu's (1986) ideas about the accumulation of 'cultural capital' prevalent within the ruling classes. The latter offers epistemic access to the language, traditions, norms and ways of thinking offered by the subject disciplines which enable young people to 'find their way', both intellectually and socially. Formal learning environments (schools, colleges, universities) are considered the most appropriate places for such theoretical concepts to be understood, in contrast to the 'everyday knowledge' gained elsewhere which largely eschews generalisation and abstraction. Concerns about whether social reproduction or social mobility is afforded through education are not far beneath the surface, questions I have briefly explored with others elsewhere (see Collins et al. 2015).

Many sociological critiques of school knowledge have focused on the relationships between knowledge and power—particularly the balance achieved between academic disciplines, school subjects and vocational education. By refusing students access to powerful knowledge, Young believes, schools actively reproduce social inequalities. As such, he argues that students are poorly served if schools construct an alternative curriculum around *their* experience:

> School may be the only opportunity that they have to acquire powerful knowledge and be able to move, intellectually at least, beyond their local and particular circumstances (Young 2009a: 15).

Young (2009b) therefore raises important concerns for all subject communities (including geographers) when he asks the following:

- In what ways is (subject) knowledge powerful knowledge?
- What aspects of (subject) knowledge do we want young people to acquire?
- How should this knowledge be organised within the school curriculum?
- How should we recognise the historical and social basis of (a subject) as an academic discipline?

Fortunately, his recent work with Johan Muller goes some way to exploring these questions through the concepts of Future 1, 2 and 3 (F 1, 2 and 3) curricula (Young and Muller 2010). In condensed form, given the space afforded for this chapter, these are explained in Fig. 2.1.

Underpinning these conceptions of 'Futures curricula' lie broader questions that Muller and Young (2008) have previously posed. For example, while recognising the crucial importance of subject-specific content in the school curriculum, they have also explored the non-arbitrariness of knowledge domains, and the connections between school and non-school knowledge. These considerations have opened up discussions, not least in the geography education community, about how knowledge is defined and the parameters within which it might be contained.

2.4 Powerful Knowledge, Geography and Geography Education

What is the connection between the theoretical conceptions of powerful knowledge, different models of curricula futures and geography education? Margaret Roberts, who debated[2] such issues with Michael Young in 2013, questions the direct applicability of the concept of powerful knowledge to geography and geography education—arguments she subsequently developed in a paper for the *Curriculum Journal* (Roberts 2014a, b). Here she pursues the interface between the theoretical basis for powerful knowledge and the practicalities of how such knowledge might be 'made flesh' in terms of both the geography curriculum and geography pedagogy in schools. Exactly what does geography teaching in schools look like if it follows the principles of promoting powerful knowledge[3]?

---

[2]Margaret Roberts and Michael Young were keynote speakers at a research seminar which considered the connections between powerful knowledge and geography education, organised by the Geography Education Research Collective (GEReCo) http://gereco.org/, at the Institute of Education, University of London in July 2013.

[3]Arguably, the debate about the place of knowledge in the school curriculum is one that has already been partially won. In 2010 the UK Coalition government's White Paper 'The Importance of Teaching' (DfE 2010) stressed the intention to move towards curricula based on 'essential knowledge'. Mitchell and Lambert (2015) refer to the recent educational policy reforms as providing an opportunity 'to engage a 'knowledge turn' with renewed focus on the role of knowledge in subject teachers' work', while the Experts' Panel (DfE 2011) for the revision of the National Curriculum expressed its support for 'giving all pupils access to powerful knowledge' (p. 11). However there is little, if any, evidence that schools have paid serious attention to these directives.

Future 1 — Boundaries are given and fixed — this 'Future' is associated with an 'under-socialised' concept of knowledge:

'Traditional knowledge' - Morgan (2014b) refers to this as the 'time honoured collection of ideas, theories, 'Great Books', and facts ... of value in their own right', observing that 'school geography is increasingly 'empty' of geographical knowledge'

Knowledge as 'given', offering a route for high achievers into the academy

Education used to introduce select social groups into dominant knowledge traditions

Transmission styles of pedagogy, 'one way' model of teaching and learning

Knowledge is static and socially conservative, continuation of the 'elite system'

'Under socialized', as it does not sufficiently recognise the social, historical and cultural conditions of its production (Morgan 2014b)

Origins in a 'system which transmits elite cultural knowledge to the 'select few'' (Young and Muller 2010, p 16)

'Treats access to knowledge as the core purpose of the curriculum and assumes that the range of subjects and the boundaries that define knowledge are largely given. It tends towards being.... a 'curriculum for compliance' and in extreme cases encourages little more than memorization and rote learning' (Young 2011a)

Future 2 — The end of boundaries — this 'Future' is associated with an 'over-socialised' concept of knowledge:

Steady weakening of knowledge boundaries

Integration of some school subjects (humanities, interdisciplinary studies)

Curriculum content understood more in terms of 'outcomes' and generic skills

'Knowledge-building' (facilitative) ways of learning favoured, rather than transmission (directive) teaching

Use of the everyday knowledge of students in the curriculum

Rise of vocational education

Socially inclusive; higher 'staying on' rates

Fig. 2.1 Concepts of Future 1, 2 and 3 curricula (after Hammond 2015)

Curriculum content and teaching methods are 'reflective of the social choices made by those who produce it and might just as easily be produced in other ways' (Morgan 2014a) Knowledge 'no longer treated as given', but 'seen as constructed in response to particular needs and interests' (Young et al 2014, p. 59) 'In its most extreme form Future 2 argues that because we have no objective way of making knowledge claims, the curriculum should be based on the learner's experiences and interests and that somehow these can be equated with the interests of society' (Young 2011a).
Future 3—Boundary maintenance observed, prior to boundary crossing. In this 'Future' it is the variable relation between F1 and F2 that is the condition for the creation and acquisition of new knowledge. Knowledge is viewed as a social product, but sanctioned by scholarly communities (with conventions, traditions, rules, etc.) which 'provide limits on what counts as knowledge' Academic communities safeguard development of disciplines/subjects in accordance with epistemic, rather than arbitrary, rules Boundary-maintenance occurs ('is this geography?') Boundary-crossing permitted, but acknowledged (Morgan 2014b) Knowledge has own status *beyond those who produce it.* Worthwhile knowledge determined by disciplinary norms (Morgan 2014b) 'Objectivity of knowledge' v 'giveness of knowledge' Curriculum must stipulate subject concepts 'that distinguish them from everyday concepts pupils bring to school' (Young 2011a). This is the starting point for curriculum construction, balancing the conceptions of Futures 1 and 2 into Future 3. Future 3 -'treats subjects as the most reliable tools we have for enabling students to acquire knowledge and make sense of the world... It implies that the curriculum must stipulate the concepts associated with different subjects and how they are related... It is this link between the concepts, contents and activities that distinguishes a Future 3 curriculum from Hirsch's lists of 'what every child should know' (Young 2014, p.67). 'What differentiates F1 from F3 (is) the induction to disciplined knowledge communities in which knowledge is not given and static but dynamic, contested and changing' (Mitchell and Lambert 2015)

Fig. 2.1 (continued)

Despite Young's insistence that powerful and 'everyday' forms of knowledge should be viewed discretely—under the contention that promoting everyday knowledge is a weak basis for developing the epistemic foundations of subject knowledge—Roberts (2014a, b) considers that everyday knowledge is in fact closely related (and important) to some themes studied in both school and academic geography. The need for students to utilise their everyday knowledge is also seen as central to the success of previous curriculum development projects in geography in the 1970s and 1980s.

Roberts (2014a, b) acknowledges Young's distinctions between 'everyday' and 'school' knowledge—the latter, in Young's view, enabling students to 'generalise beyond their experience'—and she notes his chosen focus on curriculum, rather than pedagogy. His insistence that 'school knowledge can be more abstract, more general, more systematised and go beyond what students experience in their everyday lives', enabling them to 'be more conscious of their own thinking and to have more control over it' (p. 191), is largely supported by Roberts. However, she argues that students studying geography need to bring their own knowledge, skills and understandings of the world, acquired through direct and indirect experiences (their 'personal geographies' of place, space and environment) to achieve appropriate understandings of the subject. Indeed, Roberts reminds us that most school geography curricula 'include some concepts that can be easily related to their everyday experience' (p. 192). What is also apparent within Roberts' analysis is that the study of geography and its related concepts and theories can take students far beyond such everyday experience—for example, through the introduction of concepts that are more *general* (e.g. settlements), more *abstract* (e.g. urbanization), and *beyond their direct experience* (e.g. volcanoes and earthquakes).

It is pertinent to ask, in defence of Michael Young's position, whether the geography curriculum should actively *exclude* everyday knowledge and whether this would provide a more worthwhile, intellectually fulfilling experience for students? Although Young has stated that its total removal from the school curriculum would represent an extreme reading of his views, he strongly asserts that everyday knowledge can never offer the curriculum a secure foundation. He therefore refutes claims that everyday experience provides a sound basis for curriculum construction—indeed, he states directly that the curriculum 'should not include pupil experiences' (Young 2013). Interestingly, earlier work by Hirst (1972) appears to take a 'middle line' whereby the focus of knowledge can be captured as 'experience, structured under some conceptual scheme' (p. 97). A further complication involves finding the right role for the teacher in all this. As Roberts (2014a, b) concludes:

> He (Young) assumes that teachers would make the links between everyday and school knowledge. I would argue that, on the contrary, teachers are likely to take pupils' knowledge seriously only if they are guided to do so by curriculum documents and projects (p. 194).

Essentially Roberts believes that Young raises some important issues about curriculum and pedagogy, but does not resolve them—particularly with respect to the selection of subject content to be taught. This is perhaps understandable given that all school subjects, and the disciplinary roots from which they have grown, present unique cases. The application of generic principles to subject content selection will therefore always be problematic. What concerns Roberts most, I believe, is that even the generic principles are unclear [something White (2012) also considers to be a major issue in Young's work, in his rejoinder to Young's earlier statements on students' entitlement to powerful knowledge (Young 2012)].

Maude (2015a, b), with reference to the geography national curriculum in Australia he helped to construct, opens new avenues for considering how powerful knowledge and geography education might interface. He poses three questions: (i) what might *new ways of thinking* in geography look like? (ii) what are the ways in which geographers analyse, explain and understand? and (iii) what powers do geographers have over their own knowledge? With reference to (i) he believes that geographical ways of thinking are embedded in the major geographical concepts, such as place, space, and the interconnection between people and environment. These he refers to as 'meta concepts', from which encouragement to (political) action makes the knowing 'powerful'. For (ii), analysis, explanation and understanding, he considers the concept of 'spatial distribution' to provide a helpful example—where generalisations can be used to describe processes, and analysis may be predictive. However, Maude is clear that spatial thinking is only one of many forms of geographical thinking—thinking that extends to embrace the selection of subject content, the consideration of methods used to create and test knowledge, and the identification of facts in geography. For (iii), power over our own knowledge he cites independence of thought, supported by engagement in debate and use of factual knowledge. These statements are compelling, but arguably they simply outline ways in which powerful knowledge (geographical knowledge or, in different contexts, other forms of disciplinary knowledge) helps us to think, discuss and analyse geographically.

2.5 Conclusions

Whilst acknowledging the considerable contribution to education theory made by Michael Young, and others, with respect to the role and place of powerful knowledge in the school curriculum, we may conclude that for geographers the importance of its link with everyday knowledge in the school curriculum is not yet fully articulated. Indeed, if we consider the curricula, syllabuses and schemes of work used for studying geography in English schools we see that much geographical knowledge does not currently meet Young's characteristics of 'powerful knowledge' (Major 2013; Roberts 2014a, b).

The origins of the uneasy division between every-day and disciplinary knowledge may partly lie in the readily identifiable 'gap' observed between academic and school geographies, which many believe hinder students' access to (powerful) knowledge in schools. This division has been debated recently (by, amongst others, Castree et al. 2007; Butt 2008; Butt and Collins 2013; Hill and Jones 2010) and although the dislocation between academic and school geography creates concerns, it is perhaps unsurprising that the two forms of geographical knowing and learning are largely discrete. School and academic geographies serve different purposes and meet different ends—for the key intentions of the institutions that promote them (predominantly research in the former and teaching in the latter) are different. Nonetheless, it may also be pertinent to reflect that not all the geographical knowledge that students have access to in the academy will be 'powerful'! Roberts (2014a, b) comments on these issues when she states that for many academic geographers 'everyday knowledge is a valuable resource for students, an object of study and a source of data' (p. 195) and that 'school geography, if it is to make use of the power of the academic discipline, needs to draw on ways in which academic geography uses everyday knowledge' (p. 196). Here she acknowledges Bonnett's (2008) observation that geography's 'ambition is absurdly vast' (p. 28), for what we might consider to be legitimate geographical knowledge is far reaching and not solely produced within our own disciplinary boundaries. Geography's collaboration with other disciplines, and the use of a wide range of methodologies, concepts, and theories that do not reside immediately within its academic parameters, should be noted. What is clear is that the discipline of geography must serve as the foundation from which school children understand the subject's intellectual traditions and ways of thinking, and that teachers must realise the implications of having these disciplinary roots for curriculum making and teaching in schools. Roberts (2014a, b) concludes that school geography does not always meet the criteria for powerful knowledge, but promotes *powerful ways of looking at the world* through the questions it asks, and the ways it investigates these questions.

Young, and others, have taken seriously recent criticisms by educationists about their narrow focus on theory at the expense of practice. Such criticism has particularly targeted Young's avoidance of models of subject curricula that adhere to his principles. The publication in 2014 of *Knowledge and the Future School: Curriculum and Social Justice* (Young et al. 2014) represents an attempt to visualise what the concept of powerful knowledge might look like for teachers and leadership teams in schools, whilst still maintaining at its heart a belief that social justice should be promoted through education. Here the intention is to aid thinking about the curriculum, while also encouraging the growth of teachers as 'knowledge workers' who are trusted for their expertise both in education and subject-related matters. This book was written at a time of political and educational uncertainty—towards the end of a Coalition government in the UK—when the fate of both state schools and the national curriculum in England were unclear. Nonetheless the authors present a strong vision for the future of English schools. Powerful knowledge is seen as 'a necessary component of the education of all pupils', with Lambert (2014) and Roberts (2014a, b) championing the concept of the 'knowledge-led school' that would take learners away from the narrow forms of knowledge expected of examination syllabuses, Ofsted inspections and school performance managers.

Consideration of whether geographical knowledge currently needs, to use Barnett's contentions, to be 'reinstated, reclaimed or recovered' in school geography remains open. However, many within the geography education community see an urgent need to address an apparent, widening gap between knowledge and skills. Due to curriculum centralisation school teachers have neither been encouraged to take forward their curriculum thinking, nor to engage closely with their academic disciplines. The searching observation made by Young (2011b) about the geography education community—in his contribution to a book which explored the possible futures for geography and education—is still troubling:

> What I find somewhat surprising is that although the authors of this book are as aware of and as concerned about the trends I have pointed to as I am, nowhere in these chapters do I find the powerful concepts that geography offers referred to. Is this a lack of confidence or are they taken for granted by geography educators? (p. 181).

I would contend that the geography education community must always remain diligent in (re) defining the concepts and knowledge that are appropriate for young learners—for the key curriculum question of 'what to teach?' will always exist, nuanced by educational aims, perceived student needs, political intentions, societal values, educational ideologies and the continued relevance of particular concepts. We may, as some claim, be at the point of a 'knowledge turn' in education—presaged by yet another revision of the English national curriculum—which bring 'questions of subject knowledge development to the fore' (Mitchell and Lambert 2015). There are equally pressing questions about whether schools currently promote old fashioned, or out-moded, forms of knowledge—or whether they largely ignore knowledge, due to urgent concerns about performance, pedagogy and promoting learning 'experiences'. What is apparent is that schools must ensure that they achieve the correct balance, whatever that might be, between disciplinary/powerful knowledge and the skills young people need to flourish in the 21st century. Young's concern that schools are currently poorly placed to debate 'knowledge questions'—lacking as they do any theory of knowledge—is an obvious worry; particularly as this seems to point to an inability to deliver the types of knowledge capable of reducing social inequalities and promoting life opportunities.

References

Barnett, R. (2009). Knowing and becoming in the higher education curriculum. *Studies in Higher Education, 34*(4), 429–440.

Bonnett, A. (2008). *What is geography?*. London: Sage.

Bourdieu, P. (1986). Forms of capital. In J. Richardson (Ed.), *Handbook of theory and research for the sociology of education*. New York: Greenwood.

Butt, G. (2008). Is the future secure for geography education? *Geography, 93*(3), 158–165.

Butt, G., & Collins, G. (2013). In what ways does geography progress? In D. Lambert & M. Jones (Eds.), *Debates in geography education* (pp. 230–240). London: Routledge.

Castree, N., Fuller, D., & Lambert, D. (2007). Geography without borders. *Transactions of the Institute of British Geographers, 32*, 129–132.

Collins, M., Collins, G., & Butt, G. (2015). Social mobility or social reproduction? A case study of the attainment patterns of students according to their social background and ethnicity. *Educational Review.*, *67*(2), 196–217.

DfE. (2010). The importance of teaching. London: The Stationery Office.

Firth, R. (2011). Debates about knowledge and the curriculum: Some implications for geography education. In Butt G (Ed.), *Geography, education and the future* (pp. 141–164). London: Continuum.

Hammond, L. (2015). *Can powerful pedagogies be used to support powerful knowledge in geography education*? Oxford: Presentation to GTE Conference. 31 Jan 2015.

Hartley, D. (2008). An extended review of bringing knowledge back in by Michael Young. *British Journal of Sociology of Education, 28*(6), 140–150.

Hill, J., & Jones, M. (2010). 'Joined-up geography' connecting school-level and university-level geographies. *Geography, 95*(1), 22–32.

Hirst, P. (1972). Liberal education and the nature of knowledge. In R. Peters (Ed.), *The philosophy of education* (pp. 87–111). Oxford: Oxford University Press.

Hirst, P., & Peters, R. (1970). *The logic of education.* London: Routledge and Kegan Paul.

Lambert, D. (2011). Reviewing the case for geography, and the 'Knowledge Turn' in the english national curriculum. *The Curriculum Journal, 22*(2), 243–264.

Lambert, D. (2014). Subject teachers in knowledge-led schools. In M. Young, D. Lambert, C. Roberts, & M. Roberts (Eds.), *Knowledge and the future school: Curriculum and social justice* (pp. 159–187). London: Bloomsbury.

Major, B. (2013). Geography: A powerful knowledge. Accessed at http:www.academia.edu/1327261/Geography_-_A-Powerful_Knowledge

Maude, A. (2015). *What might powerful geographical knowledge look like*? IGU CGE/UCL Geography Education Conference 'The Power of Geographical Thinking' 14 April.

Mitchell, D., & Lambert, D. (2015). Subject knowledge and teacher preparation in English secondary schools: the case of geography. *Teacher Development.* Online 15 June 2015.

Morgan, J. (2014a). Foreword. In M. Young, D. Lambert, C. Roberts, & M. Roberts (Eds.), *Knowledge and the future school: Curriculum and social justice* (pp. ix–xii). London: Bloomsbury.

Muller, J. (2000). *Reclaiming knowledge: Social theory, curriculum and education policy.* London: Routledge.

Muller, J. (2009). Forms of knowledge and curriculum coherence. *Journal of Education and Work., 22*, 205–227.

Muller, J., & Young, M. (2008). *Three scenarios for the future: Lessons from the sociology of knowledge.* DCSF and Futurelab: Beyond Current Horizons.

Maude, A. (2015). *What might powerful geographical knowledge look like*? Presentation to IGU CGE/ UCL IoE Geography Education Conference. The power of geographical thinking. 13–15 Apr 2015.

Oakeshott, M. (1972). Education: The engagement and its frustration. In R. Dearden, P. Hirst, & R. Peters (Eds.), *Education and the development of reason.* London: Routledge & Kegan Paul.

Rawling, E. (2001). *Changing the subject: The impact of national policy on school geography 1980–2000.* Sheffield: Geographical Association.

Roberts, C. (2014). Curriculum leadership and the knowledge-led school. In M. Young, D. Lambert, C. Roberts & M. Roberts (Eds.), *Knowledge and the future school: Curriculum and social justice* (pp. 139–158). London: Bloomsbury.

Roberts, M. (2011). Where's the geography? Reflections on being an external examiner. *Teaching Geography, 35*(3), 112–113.

Roberts, M. (2014b). Powerful Knowledge and geographical education. *The Curriculum Journal.*, *25*(2), 187–209.

White, J. (2012). Powerful knowledge: too weak a prop for the traditional curriculum? (May 4th) accessed at: http://www.newvisionsforeducation.org.uk/about-the-group/home/2012/05/14/powerful-knowledge-too-weak-a-prop-for-the-traditional-curriculum/

Willis, P. (1977). *Learning to Labour: How working class kids get working class jobs*. Farnborough: Saxon House.
Young, M. (1971). *Knowledge and control: New directions for the sociology of education*. London: Collier-Macmillan.
Young, M. (2008a). *Bringing knowledge back In: From social constructivism to social realism in the sociology of education*. Abingdon: Routledge.
Young, M. (2008b). From Constructivism to realism in the sociology of the curriculum. *Review of Research in Education, 32*(1), 1–28.
Young, M. (2009a). What are schools for? In H. Daniels, H. Lauder, & J. Porter (Eds.), *Knowledge, values and educational policy* (pp. 10–18). London: Routledge.
Young, M. (2009b). Alternative educational futures for a knowledge society. (May 4th) accessed at: http://socialismandeducation.wordpress.com/2009/12/06/alternative-educational-futures-for-a-knowledge-society.
Young, M. (2011b). Discussion to Part 3. In g. Butt (Ed.), *Geography, education and the future* (pp. 181–183). London: Continuum.
Young, M. (2012). The curriculum—an entitlement to powerful knowledge: A response to John White. (May 3rd) accessed at: http://www.newvisionsforeducation.org.uk/about-the-group/home/2012/05/03/the-curriculum-%e2%80%98an-entitlement-to-powerful-knowledge%e2%80%99-a-response-to-john-white.
Young, M. (2013). Overcoming the crisis in curriculum theory: A knowledge—based approach. *Journal of Curriculum Studies, 45*, 101–118.
Young, M. (2014). Knowledge, curriculum and the future school. In m. Young, D. Lambert, C. Roberts, & M. Roberts (Eds.), *Knowledge and the future school: Curriculum and social justice* (pp. 9–40). London: Bloomsbury.
Young, M., Lambert, D., Roberts, C., & Roberts, M. (2014). *Knowledge and the future school: Curriculum and social justice*. London: Bloomsbury.
Young, M., & Muller, J. (2007). Truth and truthfulness In The Sociology Of educational knowledge. *Theory and Research in Education, 5*(2), 173–201.
Young, M., & Muller, J. (2010). Educational scenarios for the future: Lessons from the sociology of knowledge. *European Journal of Education, 45*(1), Part I.

Chapter 3
Applying the Concept of Powerful Knowledge to School Geography

Alaric Maude

3.1 Introduction

The concept of powerful knowledge was introduced into educational debates nearly a decade ago by Michael Young, a British sociologist of education (Young 2008). He contends that the main purpose of schools is to teach knowledge that enables students to understand and think beyond the limits of their own experience, and describes such knowledge as 'powerful'. He argues that entitlement to this knowledge is a matter of social justice, in that all students should have access to it, and not just those from advantaged backgrounds who go to academically-oriented schools (Young 2013, p. 196). Without this knowledge young people 'are deprived and restricted in their personal and intellectual growth into fully capable adults' (Lambert 2014a, p. 1).[1] Similar arguments have been made by Leesa Wheelahan (2007) in relation to vocational education, and by Elizabeth Rata (2012) in relation to ethnic groups in the working class. Wheelahan, for example, as interpreted by Zipin et al. (2015, p. 12), argues that powerful knowledge 'provides those in adult education with critical-analytical power, beyond mere skills and facts, to read the social worlds of their practice'.

Young also claims that the concept of powerful knowledge can be used to answer the question 'what is the knowledge that we want all students to have access

[1]The extent to which some young people are deprived of access to this knowledge is illustrated by a recent study of Year 12 subject offerings in Perth [Australia] schools, disaggregated according to the socio-economic status of the school population. The study found that while all but one of the 48 schools in the top 40% by socio-economic status offered advanced maths, chemistry, physics and English literature (the subjects needed for admission to some professional university courses), only three of the 24 schools in the bottom 20% by socio-economic status offered all these subjects (Perry and Southwell 2014).

A. Maude (✉)
Flinders University, Adelaide, SA, Australia
e-mail: polymaude@ozemail.com.au

C. Brooks et al. (eds.), *The Power of Geographical Thinking*,
International Perspectives on Geographical Education,
DOI 10.1007/978-3-319-49986-4_3

to' (Young 2014a, p. 41). While geography educators have shown more interest in these ideas than those from most other disciplines, and written about them in a number of articles and book chapters (Catling and Martin 2011; Firth 2011, 2013; Morgan 2011, 2015; Roberts 2014), little has been written on what might count as powerful geographical knowledge, and therefore help answer this question for the subject. Most of the commentary by geographers has been about matters of philosophy, epistemology and pedagogy, and there are only two papers, both by David Lambert, that have any discussion of what forms of geographical knowledge might be identified as powerful. The first lists some specific examples of powerful geographical knowledge in an appendix, but without any explanation of why they are powerful (Lambert 2011). The other proposes three types of powerful geographical knowledge, but these are derived via a capabilities approach to the curriculum and not directly from the concept of powerful knowledge (Lambert 2014a).

This chapter attempts to advance the debate within geography education in two ways. The first is to develop a typology of powerful geographical knowledge that is directly derived from Michael Young's concept, while the second is to illustrate each type with examples that may make the concept clearer to geography educators. This may help them to think about the geographical knowledge students should be learning and schools should be teaching, and to answer the question posed by Young.

The methodology adopted is one of attempting to construct a logical argument. This starts with an analysis of the concept of powerful knowledge, the results of which are used to identify five types of powerful knowledge in geography. The strength of the argument is partly tested by a review of some of the criticisms of the concept of powerful knowledge by geography educators.

3.2 Powerful Knowledge

The first step in the argument is to clarify the meaning of 'powerful knowledge'. What exactly is meant by the concept, and what makes this knowledge 'powerful'? An analysis of the literature on the topic suggests that there are two ways of explaining the concept. One focuses on the characteristics that make knowledge powerful, and the other on the power this knowledge gives those who possess it.

The first theme is illustrated by these examples:

> For me powerful knowledge means, knowledge that is reliable, fallible and potentially testable … (Young 2011, p. 182).

> How can we characterise' powerful knowledge'? In short it is knowledge that is created by specialist communities or disciplines: all knowledge is a human construction, but powerful knowledge is made in accordance with some rigorous and demanding procedures and practices, put in place to test knowledge claims potentially to destruction. These state of the art epistemic practices are established to ensure that knowledge created is reliable and truthful: indeed, that it is the best it can be. (Lambert 2014a, p. 7)

These statements define powerful knowledge as knowledge that, because of its characteristics and the ways in which it has been produced within disciplinary communities, is as reliable as present understanding permits, and therefore 'powerful'. They are statements about the knowledge itself, rather than about the consequences of that knowledge.

The second way of explaining powerful knowledge describes what this knowledge can do for those who have it, as in these examples:

> *Powerful knowledge* refers to what the knowledge can do or what intellectual power it gives to those who have access to it. *Powerful knowledge* provides more reliable explanations and new ways of thinking about the world and acquiring it and can provide learners with a language for engaging in political, moral, and other kinds of debates. (Young 2008, p. 14)

> 'Powerful knowledge' is powerful because it provides the best understanding of the natural and social worlds that we have and helps us go beyond our individual experiences … (Young 2013, p. 196)

> Knowledge is 'powerful' if it predicts, if it explains, if it enables you to envisage alternatives. (Young 2014b, p. 74)

> Knowledge in the sense we are using the word in this book allows those with access to it to question it and the authority on which it is based and gain the sense of freedom and excitement that it can offer (Young 2014c, p. 20).

These two ways of describing powerful knowledge are interrelated, in that the knowledge that gives young people the powers described in the second set of statements is likely to be derived from the type of knowledge described in the first set, because that knowledge is the best available at present. However, I believe that the first way is an insufficient guide to the identification of powerful knowledge in a school subject, for two reasons. One has to do with the meaning of the word 'power'. This word generally implies an ability or capacity to do something that has an effect or outcome, so to be powerful, knowledge should have effects or outcomes that can be described as powerful. While the knowledge identified in the first way can be described as strong or robust, because of how it has been produced, we cannot assume that all of it will have powerful outcomes. Consequently another criterion is needed to guide the selection of powerful knowledge, and that is provided by the second set of statements, which shift the definition of powerful knowledge from its characteristics to what it can achieve for those who have it. Another reason for preferring the second type of explanation is that the first assumes that the geographical knowledge that schools should be teaching is disciplinary knowledge, because this is the source of knowledge that is powerful. Yet if there are significant differences between disciplinary geography and school geography, as I will be suggesting later, this assumption cannot be made. In this paper I will therefore adopt the explanations in the second set of statements, and start with the types of intellectual power that knowledge may give students.

I interpret the second set of statements as identifying knowledge as powerful if it enables young people to:

- discover new ways of thinking
- better explain and understand the natural and social worlds
- think about alternative futures and what they could do to influence them
- have some power over their own knowledge
- be able to engage in current debates of significance, and
- go beyond the limits of their personal experience.

Whether these abilities are powerful, however, is a matter of subjective judgement, because what one person regards as powerful may not be seen as such by someone else. However, while some abilities are about higher levels of understanding, which may or may not be considered to be powerful, others are about the capacity to create change, which ought to qualify as powerful.

The next step in the argument is to identify and explore the types of geographical knowledge that can give students each of these abilities.

Type 1. Knowledge that provides students with 'new ways of thinking about the world.'

Ways of thinking can be powerful because they may change a student's perceptions, values and understandings, the questions they ask and the explanations they explore. They may even change their behaviour. Geography's ways of thinking are embedded in its major concepts, such as place, space, environment and interconnection. These are not substantive concepts like 'city' or 'climate', which are about the substance of geography, but can be described as meta-concepts, which are concepts about concepts. Their role is 'to generate, at the meta-level, conceptual tools that inform the development of concepts, substantive theories and explanatory schemes, and that underpin the design of empirical studies' (Sibeon 2004, p. 13). Young himself emphasises the importance of concepts when he writes that '... intellectual development is a concept-based not a content-based or skill-based process' (Young 2010b, p. 25).

Place is a particularly rich concept, which Creswell describes as a way of 'seeing, knowing and understanding the world' (Creswell 2004, p. 11). One dimension of this can be summed up in the following statement:

> Each place is unique in its characteristics. Consequently, the outcomes of similar environmental and socioeconomic processes may vary between places, and similar problems may require different strategies in different places.[2]

This statement says that because places vary in their environmental and human characteristics, the outcomes of similar processes may differ because of their interaction with these varying characteristics. It also says that strategies to address similar problems need to take account of the distinctive characteristics of each place, which could be its environment, culture, economy, leadership or past experience. This is the core of geography's contention that 'place matters'.

[2]The statements of powerful geographical knowledge in this chapter were developed by the author, and therefore are not referenced.

Everything exists in a place, and every event happens in a place, and the characteristics of these places influence what exists and what happens. This is a fundamental part of thinking geographically, and is identified by Lambert (2014b, p. 178) as part of geography's powerful knowledge. It is powerful because it leads to questions about how to explain, and to thinking about strategies to address problems.

Another dimension of the concept of place is described in this statement:

> Places may be perceived, experienced, understood and valued differently by different people.

This teaches young people that others may perceive places differently to the way they do, because of their age, gender, ethnicity or other personal characteristics and experiences. It may help them to understand and be able to negotiate differences of opinion about planning and environmental issues, for example, which could be considered 'powerful' for both the individual and the community. It also enables teachers to integrate some of the newer areas of geography into the school subject, including the growing field of children's geographies and studies of the ways that children perceive and use places.

Environment is another rich concept in geography, and one of its dimensions can be described by this statement:

> Humans are dependent on the biophysical environment for their survival. It supports and enriches human life by providing raw materials and food, recycling and absorbing wastes, maintaining a safe habitat and being a source of enjoyment, inspiration and identity.

This summarises the four functions of the environment for people, which can be described as source, sink, service and spiritual (Maude 2014). These functions range from the practical (such as the provision of food and water) to the emotional (such as inspiring landscapes). If students adopt this way of thinking about their relationships with the environment it should influence their views on a whole variety of environmental issues. It might even change their own behaviour, or make them politically active, which could be very powerful. For example, the second function of the environment is as a sink for wastes. If students understand this function and the environmental processes involved, they may support waste management policies designed to prevent threats to human or animal health, or damage to the productive capacity of the environment. They may even adopt new waste practices, such as the safe disposal of the electronic waste they generate.

A final example of a way of thinking that is based on a geographical concept is holistic thinking, which is about recognising the interconnections between geographical phenomena. An awareness of interconnections should prompt students to think broadly and deeply in their geographical investigations, and to look for holistic and integrated explanations of phenomena. For example, to understand the causes of human-induced environmental degradation requires a study of the environmental processes producing the degradation, the human actions that have initiated these environmental processes, and the attitudinal, demographic, social, economic and political causes of these human actions (Conacher and Conacher 1995). In the case of dryland salinity students must first examine the environmental processes that produce salinity, and the human actions of vegetation removal and irrigation that caused these environmental processes. A more complete explanation of salinity, however, requires us to understand why people cleared land where there was a high risk of salinity. Was it lack of knowledge of the causes of salinity, the need to make a living, government policies or something else?

Type 2. Knowledge that provides students with powerful ways of analysing, explaining and understanding.

Michael Young argues that knowledge is powerful when it enables students to understand and explain phenomena or events, particularly those that are beyond their personal experience. There are at least three forms of geographical knowledge that can have this power. These are concepts that have analytical power, concepts that have explanatory power, and geographical generalisations.

3.2.1 Analytical Concepts

Analytical methods that can be used to identify and test relationships between phenomena are powerful because they contribute to understanding and explanation. Some of geography's analytical methods are shared with other subjects, but some are distinctively geographical because they are derived from the subject's meta-concepts. For example, the method of analysing the spatial distribution of a phenomenon for ideas on the causal processes determining its varying characteristics derives from the concept of space. A spatial analysis of a map of rainfall in Australia, for instance, shows that precipitation declines with increasing distance from the coast and rises with increasing elevation, observations which identify two of the causes of rainfall. The common method of exploring possible causal relationships by comparing spatial distributions also belongs to the concept of space.

The method of testing relationships by analysing them at different spatial scales comes from the concept of scale, and is important because different explanatory factors can be involved at different scales. For example, climate is the main determinant of the type of vegetation at the global scale but soil and drainage may be the main factors at the local scale.

Another geographical way of exploring the relationships between variables is the method of comparing places. A major report on geography in the United States argues that 'Places are natural laboratories for the study of complex relationships among processes and phenomena' (Rediscovering Geography Committee 1997, p. 30). While geographers generally cannot conduct experiments to test for relationships, they can conduct controlled comparisons of places in which one characteristic is more-or-less constant in all of them, to see whether any other characteristics of these places are also similar because of a causal relationship with the one that is constant. In teaching, the method could be used to identify the effects of a specific variable, such as climate or culture, by comparing a number of places that are similar in one of these characteristics, but different in others. For example, students could investigate the influence of climate on ways of life by selecting several places in the world with a similar climate, and finding out whether the ways of life in these places were similar because of the effects of climate, or different because of the influence of other factors.

3.2.2 Explanatory Concepts

The concept of interconnection is fundamental to explanation in geography, because causal relationships are about the connections between causes and effects. These connections involve processes or mechanisms that seek 'to show how—by what means, through which networks—particular outcomes materialize' (Gregory et al. 2009, p. 586). For example, the physical and chemical processes involved in weathering describe the mechanisms, such as freeze-thaw, that are the connection between weather and the wearing down of rock. The process of urbanisation describes the mechanisms, such as changes in the structure of the economy towards secondary, tertiary and quaternary industries, which explain why economic development results in major changes in the spatial distribution of population.

An example of a substantive concept based on interconnection is the water balance, which can be described in this statement:

> The water balance models the interconnections between precipitation, evaporation, change in water stored as soil moisture or groundwater, and runoff.[3]

This is a numerical expression of the hydrological cycle. If students understand the components of the water balance, and the interconnections between them, they will be able to use it to analyse the spatially varying availability of water as soil moisture, surface water and groundwater, and the resulting differences between places in water resources.

3.2.3 Geographical Generalisations

Generalisations are 'a synthesis of factual information that states a relationship between two or more concepts' (McKinney and Edgington 1997, pp. 78–79). Generalisation is not a specifically geographical method, but generalisations about geographical phenomena can be powerful because they help students to make sense

[3]Adapted from Davie 2008, p. 11.

of a lot of information, and so increase their understanding. They are also powerful because 'they allow students to apply what they have learned to new settings and to transfer prior knowledge to new situations' (Shiveley and Misco 2009). This enables them to ask appropriate questions and make sense of contexts beyond their experience. Geographical generalisations can be especially powerful if they include explanation or can be used to predict. This example could be developed from a study of economic geography:

> Because of the advantages of geographical concentration, economic activities tend to cluster in space unless tied to the location of natural resources or dispersed customers.

This is powerful because it synthesises our knowledge of the location of primary, secondary, tertiary and quaternary activities into one deceptively simple generalisation, which should help student understanding, and adds a major explanatory concept, that of geographical concentration. Students could use this generalisation to help explain why similar types of shops often locate together, why half the world now lives in urban areas, and why the location of employment changes as the structure of the economy changes. It can be applied to forecast the effects of anticipated changes in the structure of the economy on the future pattern of economic activity within a nation. It is also a generalisation that students may be able to challenge by finding examples that don't fit, which is likely to be an educationally and geographically valuable exercise.

Another generalisation that could be considered powerful because it includes both explanation and prediction is this example derived from physical geography:

> Because of the interconnections between the components of the biophysical environment, change in one component may produce change in others. The subsequent changes may be experienced in the same place as the initial change, and/or in different places, or at a different scale.

This is about the ways that environmental systems work, and requires a considerable knowledge of physical geography to understand and appreciate. It is also a distinctively geographical statement because it applies four meta-concepts—interconnection, change, place and scale—to produce a description unlikely to be found in a science curriculum. The result is a quite complex, yet simply expressed, guide to understanding environmental change.

There is another way in which geographical generalisations that can be used to predict are powerful, and this is that they may provide a 'basis for suggesting realistic alternatives' (Young 2010a). Students may be able to use this knowledge to forecast futures, and compare these with what they might prefer to happen. They can then think about how their preferred future could be achieved, given their understanding of the processes influencing that future. This may enable young people to identify ways of taking actions to improve their own and others' futures, an outcome which ought to qualify as powerful.

3.2.4 Criticisms of Type 2 Powerful Knowledge

Critics of the concept of powerful knowledge have particularly questioned this type of knowledge. This criticism takes several forms, only three of which can be discussed here.

One is the contention that knowledge in human geography cannot be generalised because of the influence of differences between places (Roberts 2014, p. 197), and consequently the discipline cannot produce the sort of universal truths developed in science. Young, on the other hand, argues that:

> … other forms of knowledge such as the social sciences, humanities and the arts also have concepts that take us beyond particular cases and contexts in different ways and offer some, albeit more limited and different (because of the nature of the phenomena they are concerned with) capacities for generalization (Young 2014b, p. 75).

Roberts is correct in pointing out that human geography cannot produce universal statements of knowledge similar to science, but I contend that it can still produce useful generalisations of the types described in this chapter. It is also worth noting that physical geographers also have to cope with the uniqueness of places, yet a recent paper on river systems argues that:

> Recognizing that every location is potentially unique does not render generalizations meaningless. Regularities in time and space can still be observed as repeated patterns of landforms, and interpretations of these patterns can support efforts to meaningfully transfer understandings from one location to another. The challenge lies in identifying where a general pattern holds true and how and when local differences may be important. Theory (general understanding) informs local interpretations, but site-specific appraisals prompt insights into 'differences', allowing local 'stories' to emerge (Brierley et al. 2013, p. 602; references removed)

The second criticism is based on the belief that all forms of knowledge are equally subjective, and that it is not possible to say that some knowledge is a better interpretation or explanation. The acceptance of this constructionist or relativist view of knowledge by many contemporary human geographers, and the consequent difficulty in gaining agreement on what powerful geographical knowledge might be, is well described by Morgan (2014). However, the arguments for the concept of powerful knowledge are based on a social realist view of knowledge, which rejects relativism and constructionism. Social realism contends that there is a reality that is independent of the knower, and that while our knowledge of that reality is a human construct and can never be absolute, when developed within disciplinary communities and subjected to disciplinary critique it is more reliable than an opinion or standpoint.[4]

A final criticism is that in the literature on powerful knowledge there is nothing about a moral dimension. This knowledge is considered powerful whether it is used for the benefit of others or for the personal advancement of an individual to the

---

[4]For discussions of these arguments see Firth 2013, 2015; Moore 2014.

disadvantage of others. Beck, for example, comments that 'only a little reflection shows that not all sorts of empowerment through knowledge are desirable' (Beck 2013, p. 184). Similarly, Zipin et al. (2015) argue that Young's concept of powerful knowledge overemphasises the cognitive purposes of schooling, and marginalises the ethical purposes. These are valid comments, and deserve further investigation.

Type 3. Knowledge that gives students some power over their own geographical knowledge.

The idea for this type of powerful knowledge came from thinking about this statement by Michael Young:

> Knowledge in the sense we are using the word in this book allows those with access to it to question it and the authority on which it is based and gain the sense of freedom and excitement that it can offer. (Young 2014c, p. 20)

I interpret the statement to mean that one type of powerful knowledge is knowledge that teaches students how to evaluate claims about knowledge, because this gives them the ability to be independent thinkers able to be critical of the opinions of others, including those of people in positions of power. To do this students need to know something about the ways knowledge is created, tested and evaluated within geography, and therefore about geographical reasoning. Firth makes a similar point when he argues that 'a crucial aspect of the learning of school subjects is challenging or questioning knowledge claims in the way the discipline does' (Firth 2015, p. 63). This requires students to learn the epistemic tools provided by the discipline to construct knowledge, and while some of these are generic, some are distinctive to geography and include the analytical and explanatory concepts outlined earlier.

Type 4. Knowledge that enables young people to follow and participate in debates on significant local, national and global issues.

The ability to follow and participate in public debates is essential to full and equal participation in society and its conversations about itself, and without this ability young people lack power. This is a strong justification for using geography to examine current issues, and capitalising on the subject's ability to integrate knowledge from the natural and social sciences and the humanities. This is the case in Australia, Hong Kong, Singapore and Ontario, where the geography curriculums in secondary school blend physical and human geography and use the knowledge gained to examine issues of significance to those places. The Australian secondary school geography curriculum, for example, examines issues such as water scarcity, the liveability of Australian suburbs and the management of Australian places, natural hazards, urbanisation, food security, environmental sustainability, the management of landscapes, the interrelationships between land cover change and climate change, global inequalities and migration.

Type 5. Knowledge of the world

If powerful knowledge is knowledge that takes students beyond the limits of their own experience, then the geography that teaches students about places that are beyond their experience must be regarded as powerful. This is knowledge about the world's diversity of environments, peoples, cultures and economies, which may stimulate children's curiosity, wonder and awe. It is also knowledge of their links with other places and the interconnectedness of the world, which may develop a sense of global citizenship. It is more general knowledge than that in Type 4, because it is not tied to a current event or issue.

3.3 Is Powerful Knowledge in Schools the Same as Academic Knowledge?

This chapter started with an outline of two ways of describing powerful knowledge. The first focused on the characteristics that make knowledge powerful, and implied that, to be powerful, school geographical knowledge should mostly be based on disciplinary knowledge. However, the relationship between academic disciplines and school subjects is complex (Deng 2007; Stengel 1997), and there are limits to the extent that subject knowledge can be derived from disciplinary knowledge. First, the school subject is a selection of content from the discipline, with the selection made by educational bureaucracies, curriculum writers and teachers for social, ethical, political and pedagogical reasons, not just academic ones. The selected content is then transformed by teachers into appropriate pedagogic forms for young learners (Firth 2011; Lambert 2009). As Young (2014b, p. 76) writes:

> Whereas disciplines are primarily oriented to the discovery of new knowledge (the production of new knowledge) subjects are oriented to the transmission of knowledge (the reproduction of knowledge for the next generation) and have to take account of the stage of development of learners in how knowledge content is selected (from disciplines), paced and sequenced, as well as relevance of the different theories of learning and other factors that may influence student progress.

Second, the academic subject has often moved on from the study of ideas and content that may still be relevant to school geography, as it pursues new approaches and new areas of research. For example, the concept of centrality, which is still fundamental to an understanding of settlements and the location of many economic activities, is not mentioned in a popular university human geography textbook (Daniels et al. 2012). Third, there is sometimes no consensus within the discipline that could be adopted by schools, as in the case of a major concept like scale (Moore 2008). The school subject is therefore different to the academic discipline, and while it can learn from developments within the discipline, it has to select and simplify knowledge in ways that may not mirror academic geography, and continue to teach older ideas that are still relevant to student understanding of the world. As a result some school geography may not be considered by academic geographers to provide 'the best understanding of the natural and social worlds that we have' (Young 2013, p. 196), assuming that they could agree on what that 'best understanding' might be.

3.4 Conclusion

This chapter started with the suggestion that the concept of powerful knowledge may help thinking about what we might like students to learn from their study of geography at school. Following a discussion of the meaning of powerful knowledge, the chapter identified five types of geographical knowledge, each directly derived from the concept, and described examples of each. This typology offers a more detailed classification of geographical knowledge than ones like the Geographical Association's distinction between 'vocabulary' and 'grammar' (Lambert 2011, p. 251), because it disaggregates the latter into five types. If Michael Young's argument that all students should have the opportunity to learn powerful forms of knowledge is accepted, then the typology describes ways of thinking, analysing, explaining, finding out and knowing that should be taught in school geography because they give young people intellectual powers. However, the typology does not prescribe what the content of this knowledge should be, only the forms it should take. There can be no definitive list of powerful geographical knowledge, and what should be taught must be selected by teachers to suit their circumstances.[5] Whatever topics are chosen, on the other hand, should be taught in a way that enables students to gain these intellectual powers, which will require some material to be structured hierarchically (with more complex ideas building on ones learned previously), and methods of teaching that help all students to progress well beyond factual knowledge to higher levels of thinking.

Adopting this approach to geographical education could help students to make more sense of the factual content of the curriculum, by learning how to synthesise information into generalisations or to use explanatory concepts, and to see coherence in what can often appear a somewhat disordered and sprawling discipline. In addition, the concept of powerful knowledge provides a way of explaining geography to non-geographers, by describing its ways of thinking, understanding and explaining, and demonstrating that these are powerful and involve concepts that are not taught in any other subject. Given the widespread lack of understanding of our subject in the community and amongst education administrators, this could be powerful for geography.

Several questions raised in the chapter have not been adequately answered. These include the relationship between the academic discipline and the school subject, how to identify and teach the ways that knowledge is developed and tested in geography, and whether there should be a moral dimension to powerful

[5]On the role of teachers in curriculum making see Lambert 2014b.

knowledge. A question that was not raised at all is whether the somewhat abstract ideas and concepts in powerful knowledge can be taught to all students, and if so, how. These are issues that warrant further thinking and research.

References

Beck, J. (2013). Powerful knowledge, esoteric knowledge, curriculum knowledge. *Cambridge Journal of Education, 43*, 177–193.

Brierley, G., Fryirs, K., Cullum, C., Tadaki, M., He, Q. H., & Blue, B. (2013). Reading the landscape: integrating the theory and practice of geomorphology to develop place-based understandings of river systems. *Progress in Physical Geography, 37*, 601–621.

Catling, S., & Martin, F. (2011). Contesting powerful knowledge: the primary geography curriculum as an articulation between academic and children's (ethno-) geographies. *The Curriculum Journal, 22*, 317–335.

Conacher, A., & Conacher, J. (1995). *Rural land degradation in Australia*. Melbourne: Oxford University Press.

Creswell, T. (2004). *Place: A short introduction*. Oxford: Blackwell.

Daniels, P., Bradshaw, M., Shaw, D., & Sidaway, J. (Eds.). (2012). *An introduction to human geography* (4th ed.). Harlow: Pearson.

Davie, T. (2008). *Fundamentals of hydrology*. London: Routledge.

Deng, Y. (2007). Transforming the subject matter: Examining the intellectual roots of pedagogical content knowledge. *Curriculum Inquiry, 37*, 279–295.

Firth, R. (2011). Debates about knowledge and the curriculum: Some implications for geography education. In G. Butt (Ed.), *Geography, education and the future*. London: Continuum.

Firth, R. (2013). What constitutes knowledge in geography? In D. Lambert & M. Jones (Eds.), *Debates in geography education* (pp. 59–74). London: Routledge.

Firth, R. (2015). Constructing geographical knowledge. In G. Butt (Ed.), *Masterclass in geography education: Transforming teaching and learning*. London: Bloomsbury.

Gregory, D., Johnston, R., Pratt, G., Watts, M., & Whatmore, S. (2009). *The dictionary of human geography*. Chichester: Wiley.

Lambert, D. (2009). Introduction—part 1: What is living geography? In D. Mitchell (Ed.), *Living geography: Exciting futures for teachers and students* (pp. 1–7). London: Chris Kington.

Lambert, D. (2011). Reviewing the case for geography, and the 'knowledge turn' in the english national curriculum. *Curriculum Journal, 22*, 243–264.

Lambert, D. (2014a). Curriculum thinking, "capabilities" and the place of geographical knowledge in schools. *Syakaika Kenkyu (Journal of Educational Research on Social Studies), 81*, 1–11.

Lambert, D. (2014b). Subject teachers in knowledge-led schools. In M. Young, D. Lambert, C. Roberts, & M. Roberts (Eds.), *Knowledge and the future school: Curriculum and social justice* (pp. 159–187). London: Bloomsbury Academic.

Maude, A. (2014). A sustainable view of sustainability? *Geography, 99*, part 1, 47–50.

McKinney, C., & Edgington, W. (1997). Issues related to teaching generalizations in elementary social studies. *The Social Studies, 88*(2), 78–81.

Moore, A. (2008). Rethinking scale as a geographical category: from analysis to practice. *Progress in Human Geography, 32*, 203–225.

Moore, R. (2014). Social realism and the problem of the problem of knowledge in the sociology of education. In B. Barratt & E. Rata (Eds.), *Knowledge and the future of the curriculum* (pp. 23–40). Basingstoke: Palgrave Macmillan.

Morgan, J. (2011). Knowledge and the school geography curriculum: a rough guide for teachers. *Teaching Geography, 36*, 90–92.

Morgan, J. (2014). "Neither existence nor future": The social realist challenge to school geography. In B. Barratt & E. Rata (Eds.), *Knowledge and the future of the curriculum*. Basingstoke: Palgrave Macmillan.

Morgan, J. (2015). Michael Young and the politics of the curriculum. *British Journal of Educational Studies, 63*, 5–22.

Perry, L., & Southwell, L. (2014). Access to academic curriculum in Australian secondary schools: A case study of a highly marketised education system. *Journal of Education Policy, 29*, 467–485.

Rata, E. (2012). The politics of knowledge in education. *British Educational Research Journal, 38*, 103–124.

Rediscovering Geography Committee. (1997). *Rediscovering geography: New relevance for science and society*. Washington D.C: National Academy Press.

Roberts, M. (2014). Powerful knowledge and geographical education. *The Curriculum Journal, 25*, 187–209.

Sibeon, R. (2004). *Rethinking social theory*. London: Sage.

Shiveley, J., & Misco, T. (2009). Reclaiming generalizations in social studies education. *Social Studies Research and Practice, 4*(2), 73–78.

Stengel, B. (1997). "Academic discipline" and "school subject": Contestable curricular concepts. *Journal of Curriculum Studies, 29*, 585–602.

Wheelahan, L. (2007). How competency-based training locks the working class out of powerful knowledge: a modified Bernsteinian analysis. *British Journal of Sociology of Education, 28*, 637–651.

Young, M. (2008). From constructivism to realism in the sociology of the curriculum. *Review of Research in Education, 32*, 1–32.

Young, M. (2010a). *Educational policies for a knowledge society: reflections from a sociology of knowledge perspective*. GOETE: Keynote Lecture.

Young, M. (2010b). The future of education in a knowledge society: The radical case for a subject-based curriculum. *Journal of the Pacific Circle Consortium for Education, 22*, 21–32.

Young M (2011) Discussion to Part 3. In G. Butt (Ed.), *Geography, education and the future* (pp. 181–183). London: Continuum.

Young, M. (2013). Powerful knowledge: An analytically useful concept or just a "sexy sounding term"? A response to John Beck's "Powerful knowledge, esoteric knowledge, curriculum knowledge". *Cambridge Journal of Education, 43*, 195–198.

Young, M. (2014a). Why start with the curriculum? In M. Young, D. Lambert, C. Roberts, & M. Roberts (Eds.), *Knowledge and the future school: Curriculum and social justice* (pp. 41–64). London: Bloomsbury Academic.

Young, M. (2014b). Powerful knowledge as a curriculum principle. In M. Young, D. Lambert, C. Roberts, & M. Roberts (Eds.), *Knowledge and the future school: Curriculum and social justice* (pp. 65–88). London: Bloomsbury Academic.

Young, M. (2014c). Knowledge, curriculum and the future school. In M. Young, D. Lambert, C. Roberts, & M. Roberts (Eds.), *Knowledge and the future school: Curriculum and social justice* (pp. 8–40). London: Bloomsbury Academic.

Zipin, L., Fataar, A., & Brennan, M. (2015). Can social realism do social justice? Debating the warrants for curriculum knowledge selection. *Education as Change, 19*, 9–36.

Chapter 4
Geographical Thinking: Is It a Limitation or Powerful Thinking?

Anke Uhlenwinkel

4.1 Introduction

'The Power of Geographical Thinking', is a playful variation of Michael Young's and Johan Muller's term of powerful knowledge (Young and Muller 2010). According to Young (2013) powerful knowledge is at the same time specialized and differentiated. Its specialization is articulated by it being a disciplinary knowledge that is defined by the conceptual frameworks specific knowledge communities have developed over decades and sometimes centuries. As such it is also differentiated from the everyday knowledge that students have acquired through experience and bring with them to school. In Young's view students have an entitlement to powerful knowledge as it takes them beyond what they already know.

Considering the implications that a powerful knowledge approach has for geography education, namely that teachers as professionals engage with subject content and that they cannot restrict themselves to making students happy, feeling good or training them for jobs, the German debate could not be more different from the idea of developing powerful knowledge in students. In Germany, geography is officially understood as 'an integrating subject between the natural sciences and social sciences' (DGfG 2012, p. 8), that 'makes a significant contribution to interdisciplinary and co-operative tasks in education' (ibid., p. 7). Geography as a discipline is thus defined as interdisciplinary in itself opening the door for an undifferentiated variety of contents that may be derived from all sorts of other subjects such as biology, chemistry or, less frequently, the social sciences or philosophy, but very often is simply taken from the everyday experience of the teacher

A. Uhlenwinkel (✉)
Alpen-Adria-Universität, Klagenfurt, Austria
e-mail: anke.uhlenwinkel@aau.at

C. Brooks et al. (eds.), *The Power of Geographical Thinking*,
International Perspectives on Geographical Education,
DOI 10.1007/978-3-319-49986-4_4

or from his or her 'second' or additional subject. Hence, it may not come as a surprise that currently there is no German translation for the term 'powerful knowledge'.

In the face of these clearly opposing understandings of geography as a subject the question may arise whether geographical thinking can be conceptualized as powerful thinking at all, or whether this would pose a serious limitation to the subject's aim to foster 'a considered, ethically-grounded and responsible ability to act spatially' (ibid., p. 8). To answer this question, I will first discuss the intricacies and impacts of the German geography educationists' understanding of geography as a limitless, cross-disciplinary subject, followed by a short definition of geographical thinking and then provide an example that demonstrates what the power of geographical thinking might be. In a final paragraph, I will reflect on possible limitations of this way of thinking.

4.2 Geography: A Limitless Subject?

In Britain as well as in Germany geography is repeatedly depicted as a discipline that is interwoven with an almost countless number of other subjects (for Britain: Standish 2014; for Germany: Klingsiek 2002). But while Standish emphasizes the importance of terminological rigour defining boundaries and offering classifications as well as of conceptual geographical thinking that distinguishes the discipline from others, Klingsiek's approach is firmly grounded on experience-based everyday knowledge combined with an elaborate factual knowledge (the location of Berlin cannot just be described as in the east of Germany, but needs to be seen as 50 km west of the Polish border, on the same latitude as Amsterdam etc.) and a focus on what Marsden (1997, p. 242) would call 'good causes'. Although superficially similar the two approaches are thus clearly different, when considering their conceptual foundations reflecting the debates in their respective countries. In the context of this chapter a short consideration of three positions in the German debate—an ethics-centred approach, a natural science approach, which I will mainly present as a critical comment on the ethics-centred approach, and an anti-intellectual approach—can be useful to understand the dangers of denying disciplinary boundaries.

One of these dangers has already been touched upon, which is the over-emphasis of 'good causes'. Such a development is clearly visible in the 1990s (Uhlenwinkel 2006) when for example Haubrich (1994) promoted a change in the understanding of the subject that would convert 'geo-graphy' into 'geo-ethics'. Geo-ethics in turn are neither linked to ego-centric nor socio-centric attitudes, which Haubrich both rejects as out-dated and irresponsible, but to eco-centric attitudes. According to Haubrich, ego-centric attitudes are based on striving for individual benefits and associated with capitalism and liberalism. Socio-centric attitudes are seen to consider social welfare as the main objective for ethical behaviour and eco-centric attitudes widen the scope of this view to embrace all other natural objects such as animals, plants, but also stones and soils. This ethic would strive for the maintenance of stability and an ecological equilibrium. Following Haubrich, examples of eco-centric ethics can be found in indigenous beliefs of Native American Indians, in some ecologically oriented groups within existing religions, especially Christianity and Judaism, and in Gaia meditation. Building on these ideas Haubrich considers it to be a major aim of school geography to invite students to follow the eleven commandments of ecological behaviour. Although supported by a great majority of German geography educators, Haubrich's remarks have also led to some substantial criticism, especially by Schultz (1997) and Lethmate (2000). Both of these authors discuss school geography from almost an outside view of geography criticizing not just the emphasis on the good causes that are promoted but also at least implicitly the insufficient, faulty or plainly missing subject content. In the context of this chapter Lethmate's observations and the ensuing discussion are of particular interest. The two main points that he makes are first, that Haubrich's approach comes near to indoctrination as it wishes to tell students what to do and what not to do and second, that scientific ecological thinking in school geography has no subject-specific perspective but borrows its content and its methodology from biology. This last remark highlights a perceived lack of powerful knowledge in school geography, but for Haubrich this was not the issue. Rather Lethmate's insistence on the need to determine what a subject-specific approach to ecology might entail, together with some rather unfortunate remarks on the limits behavioural genetics set to any educational effort, have prompted Haubrich to assert that he would always prefer geo-ethics to geo-determinism (Haubrich 2001). At this point the whole debate reveals an interesting constellation relating to the core of the subject content. On the one hand the juxtaposition Haubrich supposes is slightly surprising as his naturalistic ethics is based on exactly the same landscape model as that of environmental determinism, with the only difference being that in the latter people had no choice but act according to the natural conditions surrounding them while in Haubrich's model they have no choice but act according to the eco-ethics that corresponds to these natural conditions. Hence Haubrich's view could be considered deeply geographical, if traditional regional geography was still the core of the subject. On the other hand Lethmate, although concerned with powerful disciplinary knowledge, does not have a convincing concept of geographical disciplinary knowledge but remains in a realm that is still very close to biology and does not build the famous bridge between the natural and the social sciences that German school geography considers one of its key concepts. Thus one might agree with Haubrich, that Lethmate is neglecting the political aspect of geography, but Lethmate is certainly also right when he claims that Haubrich's approach is directed to changes in individual behaviour only and may be considered authoritative. In this respect Haubrich's approach is not just failing to promote powerful knowledge but it also clearly contradicts the Beutelsbacher Consensus of 1977 that identifies the limits of citizenship education as prohibiting indoctrination of any kind combined with the imperative to treat controversial issues controversially and to help students to analytically understand their societal environment and to be able to change it according to their visions (Schneider 1999). The endeavours of a considerable number of geography educationists to transmit a set of specific, often extremely naturalistic values to students instead of taking the opportunity to discuss essentially contested concepts in an appropriate way (see Tväråna 2014) have certainly weakened the subject's position in the school curriculum as they do not conform to an education for democratic citizenship.

This criticism also holds true for a different group of geography educationists that explicitly advocates an individual-centred perspective for geography lessons. This perspective corresponds to Lethmate's description of the section of young people who are more interested in living a self-complacent life than in engaging in activities to foster environmentally friendly behaviour. But although it understands itself as contrary to Haubrich's ideas of an all-encompassing responsibility, it shares with them its fundamentally anti-rational and anti-cognitive approach. Hence one of the protagonists of this perspective views maps as a collection of 'rationalistic and authoritative points and lines' (Rhode-Jüchtern 1995, p. 56) that are nothing but an expression of a 'disciplinary society' (ibid.) which hinders the motorist from finding shorter, smaller and more beautiful roads as they are not included (in a 1:1 Mio scale map). Seemingly, the author only found these smaller roads after he had forgotten his map (at the place that he found using the map) and gave himself over to the 'cunning practice of his own experience' (ibid.). Others in the same line reject the 'interpretationism' (Dickel and Hoffmann 2012, p. 12, quoting Wiesing 2009) expressed in the description of photographs when using 'rational-analytical categories and geographical patterns' (ibid.). Instead they proclaim that people are influenced by the perceptions they encounter (and not vice versa). But even if this seems to be a rather deterministic view, the authors consider it to be constructivist as people are still responsible to react to these impressions in a 'constructive', yet unreckoning way, which basically means in a way that helps them to develop their own inner selves. Now, while one may detect at least some possible geographical content in the first of these examples, what is really disturbing about the ideas in both of them is their inherently anti-intellectual perspective that not only neglects their own discipline, but any kind of rational discourse that forms part of a knowledge-society, and for that matter also of powerful knowledge. Especially because of this they also have the potential of significantly weakening the subject's position in the curriculum.

All three strains of the German geographical education debate, Haubrich's geo-ethics, Lethmate's natural science approach as well as Rhode-Jüchtern's anti-intellectual perspective, are good arguments for limiting the discipline to an identifiable, even if always contested core. Whatever their merits may be in the eyes of their advocates, there is one thing that none of these approaches can provide and that is powerful geographical knowledge. Nonetheless I will here concentrate on the overtly anti-rational approaches as they are hardly suitable to 'stress the emancipatory power and purpose of education in initiating young people into forms and fields of specialized knowledge—without which they are deprived and restricted in their personal and intellectual growth into fully capable adults' (Lambert 2014, p. 13).

4.3 Geographical Thinking

One of the reasons why German (school) geography is in such a comparatively miserable state is the lack of any conceptual approach through which geography might be identified as a subject in its own right (Uhlenwinkel 2014). In the national Educational Standards geography is defined as being concerned with 'geographically and geo-scientifically relevant phenomena and processes such as globalisation, climatic change, earthquakes, flooding and storms, as well as population change, migration, disparities and conflicts over resources, (that) shape many aspects of our lives and our societies on planet Earth' (DGfG 2012, p. 5), thus by the objects studied rather than in relation to any geographical concepts that might help to identify the limits of the subject.

The rather essentialist philosophy underlying this subject understanding contrasts with debates in Britain, France and the United States, where geography is defined by a number of key concepts (Uhlenwinkel 2013) either with a focus on spatial thinking or with a focus on thinking geographically. The American spatial thinking approach, that relates back to the works of some of the better known representatives of German academic geography, such as Christaller and his Central Place Theory, is grounded in abstract mathematical models. In the context of this approach space is defined by properties such as dimensionality, continuity, proximity and separation, all of which make it measurable in terms of kilometres or miles, time distances or with reference to different coordinate systems, ideas that would clearly be dismissed as rationalistic by several German geography educators including Rhode-Jüchtern (see above).

The geographical thinking approach that can be found in different versions in the UK and in France denotes a decidedly subject-based idea. People who think geographically use geographical concepts such as place, space, scale, interdependence, diversity, proximity or distance to analyse the phenomena in the world around them. The meanings as well as the choice of the key and organizing concepts are constantly changing (Taylor 2008) and can also differ between authors. Despite the use of partly different terms and the diversity of compilations there are a number of similarities though, such as the focus on place, space and often also scale as key concepts. These in turn are related to notions such as meaning-making and power geometries which mark the subject as a human or social science and link it to wider philosophical debates without duplicating them. Although geographical thinking thus offers a broader access to geography topics in school than a factual knowledge based approach does, German geography educators in the tradition of Haubrich's approach would prefer not to take the subject content detour.

Hence, from a German perspective both spatial thinking and thinking geographically must seem like an almost unbearable limitation to moralistic demands and individual self-expression. From a rationalist or even cognitive view though, it may be the geo-ethics and individual-centred approaches that may be felt as an unjustifiable limitation to human understanding of the world. In the following Sect. 4.4 I will show how geographical concepts can be used in argumentation to produce some powerful thinking that will help students to rationally understand situations and relate this understanding to discourses on essentially contested concepts.

4.4 Powerful Thinking in the Geography Classroom

In the academic discourse on the work of Immanuel Kant there is a debate on whether in his philosophy the law drives more or less directly from ethical principles or whether these two domains are more or less independent of each other (Horn 2014). One argument for their independence is that ethical decisions have to be made by autonomous individuals who believe they are doing something right, while legal regulations only demand a certain behaviour, whether people agree to it or not. This difference may be one of the reasons why legal arguments are a lot easier to decide than ethical ones (Alexy 2012 [1978]). Analogously it can be said, that geographical rules cannot be derived from ethical principles and that geographically described and explained conflicts would be a lot easier to understand, if questions of values were not involved.

Just as the law is limited to a certain type of norms, so are the rules of the legal argumentation different from those of a general argumentation. According to Alexy (ibid.) general rules of argumentation would for example encompass the demand that no speaker may contradict himself or that every speaker has to be able to give reasons for his claims. Examples of some of the specific rules of legal argumentation would be the requirement that each legal statement has to include at least one legal norm or that the proponent has to take as many argumentative steps as it takes to reach concepts that are indisputable in relation to a given case.

Transferring this theoretical approach to the discipline of geography what would be needed is specific rules of geographical argumentation as the general rules would apply just as in the case of a legal argument. Modelling these on the examples just given, I suggest that one such specific rule could be that each geographical argument has to include at least one reference to a geographical concept or that the geographical arguments have to take into account as many scales or layers as are needed to understand a specific place or space.

To discuss the merits of such an approach I will use the example of wind power production as it seems to be a recurring topic in school geography in several of the countries considered here and as there are also substantial political debates pertaining to it. I will start with two non-geographic argumentations.

In the political debates in almost every country, opposing voices can be heard which although asserting that 'whether forests of wind turbines are a vision of beauty or an environmental outrage is a matter of personal opinion' (Lawson 2008, p. 69) clearly wish to convey that the latter is the case, and thus in an almost romantic manner claim that 'les amoureux de l'environnement s'en mêlent et s'insurgent contre l'installation des fermes éoliennes sur des sites emblématiques ou dans des parcs naturels régionaux' [the lovers of the environment interfere with and stand up against the construction of wind farms on emblematic sites or in regional nature parks] (Cabrol 2010, p. 104). No matter whether one agrees with these viewpoints or not, they are clearly not geographical arguments. Rather they might fit into the category of individual-centred perspectives that can be found in the German educationists' debate.

On the part of other environmentalists the number of birds killed by wind turbines causes some concern. Studies in Spain show that with regard to the deaths of griffon vultures the wind speed in the parks may be of crucial importance, with lower and medium wind speeds leading to more casualties than high wind speeds (Barrios and Rodríguez 2009, p. 250). Even though these findings are certainly interesting the question has to be asked whether the mere fact that the data were collected in distinctive locations, two wind farms five and ten kilometres north of Tarifa, makes the results geographical. In the light of the two examples of specific rules for geographical argumentation one would certainly have to argue against this supposition, as there is neither a reference to a geographical concept nor have different scales or layers been taken account of. This example may satisfy criteria of measurability or scientific research, but without an appropriate contextualisation it cannot (yet) be viewed as a geographical argument.

4.4.1 An Argument Using the Concept Space

To construct a geographical argument according to the examples of specific rules suggested above, what is needed first is a reference to a geographical concept such as space. The concept of space describes the interrelations between different places, and thus cannot be conceived without the concept of place (Massey 2005) that will be focussed on in the next section. Places are constantly changing and they also change each other through their developments. Hence spaces are not fixed, but only stories-so-far. The German pattern of wind power production tells such a story that has recently led to a heated debate in the political realm.

In 2005 Germany was the largest producer of wind energy worldwide followed by Spain and the United States of America (Denhez 2009). A map of the distribution of installed renewable energy capacities shows that most of the wind farms needed to achieve this can be found in the northern federal states (Deshaies 2011). To an observer with strong inclinations towards environmental determinism this may not seem astonishing as the highest regional wind speeds in the country are found along the coast and in the northern parts generally, although this 'north' reaches as far south as Frankfurt (Barré and Mérenne-Schoumaker 2011), which means that it includes most of the federal state of Hesse. In this federal state it is not wind speeds that are inhibiting the development of wind energy capacities, but a restrictive legislation that confines possible construction sites to an extremely small percentage of the surface area (Scheer 2008). Unfortunately though, those federal states that most effectively block the expansion of wind farms are also among those with the highest demand for energy as they host two of the three main industrial centres of the country (Deshaies 2011).

The inequality in the distribution of production of and demand for energy has led to plans to build three main power lines from the north to the south. One of the beneficiaries of the lines would be Bavaria. But power lines would be just as much of 'an environmental outrage' (Lawson 2008: 69, see above) as are wind turbines to some. Consequently there are substantial regional protests against the power lines which have lately been addressed by a Bavarian politician who proposed building the power lines around the federal state instead of through it (StMWi 2015).

The geographical argument here is clear, but is the proposal of the Bavarian politician responsible or justified? And can or should geographers give definite answers to the ethical questions involved?

4.4.2 An Argument Using the Concept Place

Depending on the problem discussed other concepts may appear more suitable. The following example uses a case study from Spain and focusses on the concept of place. Cresswell (2004) identifies three criteria to define place: it has to have a location, a material setting and a meaning that people attribute to it. The commons of medieval Spanish villages for example can be conceptualised as places. They have a location, usually on hilltops somewhere in the area that belongs to the community. They have a certain material setting as they were not used for growing crops, but were either a source of wood that has now been long exploited or were used for herding. All incomes that were made from these lands belonged to the community. Hence the commons also carried a specific meaning for the villagers.

Today most rural communities in Spain face a decrease of their population because younger people move to the bigger cities to find work. This first created a situation where older people were left behind, with no services and amenities, as the communities could not pay for them for lack of income through taxes. In recent years though, some communities have seen a more positive development: They have reduced their public debts to zero and invested in the restoration of their churches, ancient wells and communal ovens. They also installed water pipes conducting water from the mountains to the villages and meeting places for the villagers. Apart from that, some have even created new employment for social workers. These developments have been facilitated by lease charges, which are received from companies operating wind parks on the former commons (Bartmann 2008). The reason why the communities profit from the installation of wind parks can be found in traditional land rights, as in a lot of communities the *juntas vecinales* are still the owners of the commons. The income from these lands has to be invested in either the mountains or purposes of communal interest, which leads to the development of these communities. As such, the argument may be one of empowerment. But it becomes more complex when the griffon vultures come in.

Vultures live in cliffs and feed on dead animals that they find on the ground. Both these features imply that they are very likely to use the same areas as people use to install wind turbines. This in itself would be of little significance if the vultures simply passed the wind turbines in flight. But the study quoted above has also shown that most collisions of the birds with the wind turbines occur during take-off when the vultures use a circling flight mode to gain height (Barrios and Rodríguez 2009). Taking into account that regionally speaking in Spain the areas of high wind energy production and the areas that are the feeding grounds for vultures largely overlap (Iñigo and Atienza 2007; Atienza 2009), we here have a conflict over place, that includes the question of power geometries (Massey 1999) relating to villagers versus most probably urban NGOs of bird lovers.

Here again, the geographical argument is quite easy to make, but the value judgements involved are less easy to solve.

4.4.3 An Argument Using the Concepts of Place, Space and Scale

While in Spain the building of wind turbines has positive effects on the communities the same process is met with scepticism in Mexico. From a geographical perspective two reasons can be identified: One is concerning the concept of place, the other one the concept of scale.

The Istmo de Tehuantepec is one of Mexico's windiest regions and thus viewed as a perfect place to build wind turbines. It is also a region with a large indigenous population. To strengthen indigenous land rights the Mexican constitution defined three different kinds of land property: private property, community property and *ejido*-property. Community land is meant to be used for communal purposes such as growing and maintaining forests or the construction of infrastructure. *Ejido* land is given to landless people to give them the opportunity to grow their own food (Lehmann 2014). This right to land has been restricted by the constitutional reform of 1992 that facilitates privatisation of *ejido*-land (Gerber 2013). Also since 1992 the administration of the communal land can be based either on a democratic system with parties being voted into power or on a traditional system (*uso y costumbres*) that favours local celebrities (Owolabi 2004). This system has led to the rather paradoxical result that the party that has once supported the designation of community and ejido-property, now uses the loyalty of local indigenous leaders to make the land available to investors who want to build wind farms. Often the lease charges have thus been privatised as well (Gerber 2013). Both these developments have led to a situation where local people may be deprived of their rights when wind parks are planned and build. For the constitutional reform this is obvious, as it becomes easier to sell and buy land formerly used by indigenous people. The dualism of two possible systems of local governance has led not only to conflicts over the system being used, but also to decisions being made by the respective representative without consultation of the community. Consequently wind farms in Mexico have a completely different status and significance from what they have in Spain.

But although the construction of wind turbines is part of quite different stories-so-far in the two places, it is also intimately interlinked, which leads to an argument using the concept of space. The starting point for this linkage was the artificial shortage of land resources and of access to the power distribution network by the Spanish government. This resulted in a competition between Spanish energy producers for sites which the regional governments would only grant if the companies guaranteed extra investments in their regions. These extra investments properly excluded foreign firms from the market thus having the effect of protectionist measures that helped Spanish firms to grow (Bartmann 2008). Once grown to an internationally competitive scale, these firms are now trying to expand to the Mexican market, using resources from national and international banks as well as international organizations such as the Clean Development Mechanism or the Clean Technology Fund to further their business (Lehmann 2014).

The ideological surplus that the latter organizations offer can be endorsed by an extended concept of scale that goes beyond the usual metric measurements and includes the meanings we contribute to different scales (Lambert and Morgan 2010). In this case, the meaning produced on a global level is that of fighting climate change and poverty, while supporting sustainable development in less privileged regions of the world.

As in the two preceding examples the geographical argument is readily accessible, but how do we judge these developments? Which would be responsible and just decisions to take? Who should benefit? And who should pay the price?

4.5 Limitations as a Way to Strengthen Decision Making Capabilities

In the above examples the geographical arguments can easily be understood by everyone: Bavarians need a power supply, but they want neither windmills nor power lines, which makes it hard to provide them with renewable energy. Spanish villagers benefit from wind power installations, but at the same time, by helping companies grow, jeopardize the living conditions of Mexican villagers. You can add in power geometries that may make the situation more complex but still comprehensible: Bavarians have the money, while federal states in northern Germany need to generate economic growth. In these federal states, the people who profit from the wind mills, be it energy producers or farmers renting the land, have other options than people complaining about the destruction of natural habitats, unless the latter succeed in forming a strong pressure group. In the Spanish-Mexican example similar power-geometries can be discerned. The whole argument becomes a lot more difficult to decide when value questions come in: Is it right to protect Bavarian landscapes but not those in northern Germany? Is it alright to build a day care centre in rural Spain when this means ruining communities in rural Mexico? Should people be treated equally or are their circumstances so different that they should be treated differently as well? These questions are all important and they should certainly have a place in school subjects that play a larger part in citizenship education; they need not only be discussed on a firm basis of powerful disciplinary knowledge in geography but also in philosophy or political science. In this context differences between universal, personal and polysemantic values have to be taken into account just as much as different levels of civic reasoning—such as different definitions of the concepts in question, analysis of the reasons for different opinions relating to the concepts and analysis of the basis of different understandings of the principles (Tväråna 2014).

4.6 Conclusion

Is geographical thinking then a limitation or powerful thinking? It may certainly be seen as a limitation when the issue in question is not one that can easily be solved by geographers (alone), such as ethical questions or questions of self-realization. But when it comes to understanding situations that involve regional disparities and conflicts geographical thinking is not a limitation, but opens ways of understanding that moral instruction alone cannot offer.

Nonetheless, especially in a school context, where geography is seen as one of the subjects most responsible for citizenship education, value questions cannot be neglected either. But teachers should treat these aspects as powerful knowledge as well, meaning that they will have to take into account the differences between value systems (utilitarian, naturalistic, intuitionist, rational) and the different levels of reasoning involved so that students can see the principles behind certain understandings of the values in question. If teachers feel that this is asking too much, then at the very least they should be a lot more cautious about the limits between their own subject and others instead of trying to teach geography in a way that neither serves geography nor gives attention to the values dimension of geographical issues.

References

Alexy, R. (2012). *Theorie der juristischen Argumentation. Die Theorie des rationalen Diskurses als Theorie der juristischen Begründung*. Frankfurt am Main: Suhrkamp.

Atienza, L. (2009). *Wind energy development in Spain*. Retrieved from http://de.slideshare.net/JoeloRoss/mit-wind-power-development-in-spain-2009-9832983

Barré, B., & Mérenne-Schoumaker, B. (2011). *Atlas des énergies mondiales. Un développement équitable et propre est-il possible?*. Paris: Éditions Autrement.

Barrios, L., & Rodrígues, A. (2009). Patrones espaciales y temporales de mortalidad de aves en dos parques eólicos del sur de España. In M. de Lucas, G. Janss, & M. Ferrer (Eds.), *Aves y parques eólicos. Valoración del riesgo y atenuantes* (pp. 241–253). Madrid: Quercus.

Bartmann, C. (2008). *Windenergieanlagen – Entwicklungsimpulse für ländlich periphere Regionen in Spanien? Entwicklungseffekte von Windparks in drei ländlich peripher gelegenen Kommunen in Spanien*. Unpublished master thesis, Potsdam.

Cabrol, L. (2010). *En vert et contra tous*. Paris: Le cherche midi.

Cresswell, T. (2004). *Place—A short introduction*. Oxford: Blackwell.

Denhez, F. (2009). *Atlas du changement climatique. Du global au local, changer les comportements*. Paris: Éditions Autrement.

Deshaies, M. (2011). *Atlas de l'Allemagne. Les contrastes d'une puissance en mutation*. Paris: Éditions Autrement.

DGfG, (2012). *Educational standards in geography for the intermediate school certificate. With sample assignments*. Bonn: Selbstverlag.

Dickel, M., & Hoffmann, K. (2012). Mit Bildern umgehen – Zwischen Spielraum und Festlegung. *Geographie und Schule, 34*(199), 12–19.

Gerber, P. (2013). *Europäische Unternehmen erzwingen das grüne Geschäft mit dem Wind in kolonialem Stil*. Conquista 3.0: Die Windkraftindustrie in Oaxaca, Mexiko. Retrieved from http://www.rosalux.de/fileadmin/rls_uploads/pdfs/Standpunkte/Standpunkte_international/Standpunkte_int_07-2013.pdf

Haubrich, H. (1994). Globale Aspekte der geographischen Erziehung. In M. Flath & G. Fuchs (Eds.), *Die Erde bewahren—Fremdartigkeit verstehen und respektieren* (pp. 52–63). Klett: Gotha.

Haubrich, H. (2001). Bio-Ethik versus Bio-Determinismus. Zur Reaktion von Jürgen Lethmate (GUID 2001, H. 1. S. 21ff.) auf meine Kritik (GR 52, H. 10) S. 61f.) an seinem Beitrag „Das geoökologische Defizit… (GR 52, H. 6, S. 34 ff). *Geographie und ihre Didaktik, 29*(2), 99–102.

Horn, C. (2014). *Nichtideale Normativität. Ein neuer Blick auf Kants politische Philosophie*. Berlin: Suhrkamp.

Iñigo, A., & Atienza, J. (2007). *Efectos del Reglamento 1774/2002 y las decisiones adoptadas por la Comisión Europea en 2003 y 2005 sobre las aves necrófagas en la península Ibérica y sus posibles soluciones*. Retrieved from http://www.seo.org/media/docs/Informe%20carro%F1eras.pdf

Klingsiek, G. (2002). Warum ich gerne Erdkunde unterrichte. *Geographie Heute, 23*(200), 2–3.

Lambert, D. (2014). Curriculum thinking, 'capabilities' and the place of geographical knowledge in schools. *Prace komisji edukacji geograficznej, 3*, 13–30.

Lambert, D., & Morgan, J. (2010) *Teaching geography 11–18. A conceptual approach*. Maidenhead: Open University Press.

Lawson, N. (2008). *An appeal to reason. A cool look at global warming*. London: Duckworth Overlook.

Lehmann, R. (2014). Mehr als nur (heiße) Luft. Windparks im mexikanischen Bundestaat Oaxaca verfestigen Ungleichheiten. *Peripherie, 136*(34), 427–444.

Lethmate, J. (2000). Das geoökologische Defizit der Geographiedidaktik. *Geographische Rundschau, 52*(6), 34–40.

Marsden, B. (1997). On taking geography out of geographical education. *Geography, 82*(3), 241–252.

Massey, D. (1999). *Power-geometries and the politics of space-time*. Heidelberg: Department of Geography.

Massey, D. (2005). *For space*. London: Sage.

Owolabi, K. (2004). ¿La legalización de los "usos y costumbres" ha contribuido a la permanencia del gobierno priista en Oaxaca? Análisis des las elecciones para diputados y gobernadores, de 1992 a 2001. *Foro Internacional, 177*(3), 474–508.

Rhode-Jüchtern, T. (1995). *Raum als Text. Perspektiven einer Konstruktiven Erdkunde*. Wien: Materialien zur Didaktik der Geographie und Wirtschaftskunde.

Scheer, H. (2008). Die Koalition der Verhinderer und ihre Methoden. Wie administrative Hemmnisse den Boom der erneuerbaren Energien bremsen. In S. Kosch (Ed.), *Zukunftsmarkt Klimaschutz. Wie wir die Welt retten und dabei Geld verdienen können*. Berlin: Parthas Verlag, pp. 112–122.

Schneider, H. (1999). Der Beutelsbacher Konsens. In Mickel, W. W. (Ed.), *Handbuch zur politischen Bildung. Grundlagen, Methoden, Aktionsformen*. Schwalbach/Ts.: Wochenschau Verlag, pp. 171–178.

Schultz, H.-D. (1997). Mit oder gegen die Natur? Die Natur ist, was sie ist, und sonst gar nichts. *Zeitschrift für den Erdkundeunterricht, 49*(7–8), 296–302.

Standish, A. (2014). Some important distinctions for geography educators. *Geography, 99*(2), 83–89.

StMWi. (2015). *Stellungnahme zum Netzentwicklungsplan. Aigner: "Faire Lastenverteilung bei neuen Stromtrassen"*. Retrieved from www.stmwi.bayern.de/presse/pressemitteilungen/pressemitteilung/pm/102-2015/

Taylor, L. (2008). Key concepts and medium term planning. *Teaching Geography, 33*(2), 50–54.

Tväråna, M. (2014). *Rikare resonemang om rättvisa. Vad kan kvalificera deltagande i samhällskunskapspraktiken?* Stockholm: Stockholm University.

Uhlenwinkel, A. (2006). *Blick zurück nach vorn. Zur Entwicklung der Geographiedidaktik in Deutschland seit 1969*. Unpublished habilitation thesis, Bremen.

Uhlenwinkel, A. (2013). Spatial thinking or thinking geographically? on the importance of avoiding maps without meaning. In T. Jekel, A. Car, J. Strobl G. Griesebner (eds.), *GI_Forum 2013. Creating the GI Society. Conference Proceedings* (pp. 294–305). Berlin, Wien: Wichmann.

Uhlenwinkel, A. (2014). Factual knowledge and conceptual understanding. *Geography, 99*(1), 28–35.

Wiesing, L. (2009). *Das Mich der Wahrnehmung*. Frankfurt am Main: Suhrkamp.

Young, M. (2013). Overcoming the crisis in curriculum theory: A knowledge-based approach. *Journal of Curriculum Studies, 45*(2), 101–118.

Young, M., & Muller, J. (2010). Three Educational Scenarios for the Future: lessons from the sociology of knowledge. *European Journal of Education, 45*(1), 11–27.

Part II
Pedagogy and Geographical Thinking

In a book entitled *The Power of Geographical Thinking*, it will come as no surprise that the editors see significance in the ways in which this phenomenon can be applied in education. As a result, in this part, the seven contributing authors focus on the ways in which teachers and teacher educators can enhance powerful geographical thinking specifically through pedagogy. There are a number of common denominators here, beyond the binding thread of geographical thinking and pedagogy. The first two chapters examine the significance of teachers' critical thinking about powerful geography theories and concepts. The next three chapters examine ways in which geography subject knowledge can be enhanced through teachers developing pedagogies for powerful geographical thinking. The topics that authors have chosen to discuss are likely to strike a chord with contemporary geography teachers and geography teacher educators. In particular, there is a strong environmental thread, with an emphasis on teaching about sustainable futures, the complexities of climate change and developing students' reflective thinking on the human/environment relationship. The final two chapters focus on another key area of development in geography teaching and learning—the role of spatial thinking and geographic information systems in geographical thinking. What each of these chapters has in common is the unshakeable conviction that developing the power of geographical thinking lies at the core of high quality geography education.

To begin, Janis Fogele's chapter, *Acquiring Powerful Thinking Through Geography Key Concepts* provides a useful link between the discussions already presented in part 1 on powerful thinking with geographical concepts. She focuses on the role of teachers in using specific geography subject knowledge concepts as 'leitmotifs' to engage their pupils. Her argument hinges on the significance of teaching pupils about human–environment interaction through enquiry questions formulated around key concepts, such as space and place.

In *What type of geography do we teach? From theoretical-conceptual weaknesses to underestimation of spatial experience. Chilean teachers' views on teaching geography*, Andoni Arenas-Martija, Patrico Peréz-Gallardo, Victor Salinas-Silva and María José Otero-Auristondo Salinas-Silva, Arenas-Martija and Ramirez-Lira's chapter provides a valuable discussion on the connections between a teacher's individual subject knowledge and the ways in which they develop powerful geographical thinking in their students.

In *Geographical thinking and its role in climate change education: A case of Singapore*, Chew Hung Chang argues that deeper understanding of climate change requires full engagement with the power of thinking geographically. He examines the challenges that teachers face in teaching about climate change in Singapore. Research findings identify both pupils' misconceptions and naïve knowledge about climate change and the challenges teachers face in engaging pupils with a phenomenon that they often know very little about. He concludes that to grasp the complexity of climate change young people need to be taught about both human–environment interactions and other key geographical concepts. These include space, place, scale and movement. He uses Young and Muller's F3 curriculum model (2010) to argue that these are unique constitutents of the powerful nature of geographical thinking.

In their chapter, *Geographic Education for Sustainability: Developing a bi-national geographical thinking curriculum*, Fabian Aria Palacios, Alex Oberle, Imena Cortés Quezada and Mollie Ullestad discuss ways in which teachers can develop powerful geographical thinking about sustainability. They argue that the key concept of sustainability can be used to link local geographical issues with the wider global context. In connection with the main theme of this book, and drawing on Peter Jackson's definition of thinking geographically, the authors identify strongly with the idea that geographical thinking offers unique perspectives with which to view the world. In particular, they argue that teaching about interdependence and the links between the physical and human worlds can encourage an understanding of sustainable development.

In her chapter, *Students' reflective thinking about geographical content: Using weblogs to determine the levels of student reflection in global education* Nina Brendel considers the pedagogical role of weblogs in students' reflective geographical thinking. She considers specifically the role of reflecting on geographical content and how this allows students to access higher levels of reasoning. The author uses case studies of particular ways of geographical thinking, illustrating, for example, how developing an appreciation of a systems approach can deepen geographical understanding.

In his chapter, *Geographical and Spatial Thinking in the Swedish Curriculum*, David Orbring examines the connections between geographical and spatial thinking in the Swedish curriculum. He focuses on knowledge in the Swedish curriculum and places this in an international perspective, first explaining how specific geographic abilities are defined in the Swedish context and then discussing how these are taught through specific pedagogies in schools. These abilities are considered both with regards to geographical thinking and with reference to one particular type of geographical thinking—spatial thinking. The chapter draws on the preliminary stages of a research project which considers an analysis of the Swedish geography curriculum. The chapter describes abilities in geography and critically examines what teaching in geography (from a Swedish point of view) should include.

The discussion concludes by drawing our attention to the crucial significance of geography teacher expertise in developing pupils' abilities to think geographically.

In her chapter, *GIS and the power of geographical thinking*, Mary Fargher examines the role of GIS in developing geographical thinking. She examines the role of the teacher in constructing geographical knowledge through GIS and what this means for their pedagogical approach, particularly with regards to enquiry learning. Her central argument is that teachers need to be more critically aware of the kinds of geographical thinking that can and cannot be enhanced through GIS. She identifies the knowledge limits of GIS, drawing on the broader academic field of critical GIS, an area previously under-explored in school geography education. The chapter concludes by reviewing the role of new web GIS and the opportunities and challenges that it brings for teachers for developing geographical thinking.

In conclusion, this part considers the 'whys' and 'hows' of geography teachers applying the principles of powerful geographical thinking to pedagogy. From geography teacher education on geographical enquiry, based around key geographical concepts (Fogele), to using Young's powerful knowledge concept in addressing pupils' misconceptions on climate change (Chang), the chapters in this part highlight the significance of developing teacher expertise in extending pedagogies which can enhance powerful geographical thinking.

Chapter 5
Acquiring Powerful Thinking Through Geographical Key Concepts

Janis Fögele

5.1 Introduction

There is a growing awareness of the importance of improving geographical thinking in school geography (Lambert 2013). To fully understand some of the most recent major world issues—such as climate change, the refugee problem, migration and globalisation—we need to focus on the interconnections between human, environmental and physical factors at different scales. In short: a geographical view on these problems can help to cope with their complexity (Morgan 2012).

This leads us to recognise an increasing demand to support geographical thinking in schools (Jackson 2006), instead of an additive learning of places and subject-specific terms. Using geographical key concepts to organise and intertwine content in geography lessons is a means of creating progress concerning the students' cumulative learning and their geographical thinking skills (Bennett 2005). Key concepts can be defined as subject-specific ideas or analytical frameworks emerging from the structure of a subject to describe the most relevant processes and phenomena (Uphues 2013). This supports an 'understanding of a conceptual framework that helps students to see geography as an intellectual body of knowledge rather than as a set of disparate units of study' (Walshe 2007, p. 101).

For the implementation of these ideas we have to focus on teachers and their understanding of teaching and learning in geography classes, their knowledge about the subject and about ways of teaching that enable students to develop their thinking. Hence this chapter focuses on the potential of key concepts for teachers' work: How teachers can be supported in developing their capacity to think geographically, and how they can acquire subject-specific thinking skills that are powerful—as a means of empowering their capacity, and that of their students, to deal with complex issues in a very analytical, systematic and confident way. This is

J. Fögele (✉)
University of Giessen, Giessen, Germany
e-mail: Janis.Foegele@geogr.uni-giessen.de

C. Brooks et al. (eds.), *The Power of Geographical Thinking*,
International Perspectives on Geographical Education,
DOI 10.1007/978-3-319-49986-4_5

crucial bearing in mind recent issues like the global food supply crisis, or the struggle for finite resources, that are of great interest for geography classes. Therefore 'geography teachers need to re-engage with their subject matter to enable them to improve how they teach geography' (Brooks 2006, p. 353). This diagnosis leads to the need to give some thought to the potential and challenge of the key concept approach.

This chapter starts by providing a short introduction to the key concept approach in geography education. Then it turns its attention to one key concept approach that is at the core of the German-speaking educational standards (Curriculum 2000+ and partially in DGfG 2012, 13): four different *concepts of space* are presented, supplemented by a short overview of some further concepts such as *structure|function|process* as parts of the main concept *system* (DGfG 2012, 11). Subsequently some considerations are shared, enabling teachers to use this approach for their lesson planning to promote geographical thinking. The importance of providing challenging exercises to enable a cognitively activating geographical enquiry is emphasised. Based on two exemplar exercises, an idea for the implementation of key concepts and challenging enquiry exercises demonstrates how these approaches can offer a chance to acquire powerful thinking, both for geography teachers and students.

The chapter reflects parts of a research project which aimed at learning about the implementation of key concepts in geography classes and discovering changes in the teachers' thoughts about teaching geography through this educational and methodical approach. The study was based on a six-month in-service training package for geography teachers on geographical key concepts (see Fögele and Mehren 2015). Here I do not present a fixed research design, but focus on what has been learnt during the training of geography teachers concerning the needs of a practical implementation of key concepts in classrooms. The concluding section offers some suggestions for the development of in-service training about key concepts in geography.

5.2 Geographical Key Concepts

A definition of key concepts understands them to be 'the structured interconnection of related concepts, theories and explanatory models that have emerged from the structure of a subject for the description of elementary processes and phenomena' (Demuth et al. 2005, p. 57). They occur in the consideration of a number of (global) issues in geography lessons, and in everyday life, and can therefore contribute to the structured interconnection of topics. As long as the key concepts are not additional knowledge concepts they serve those learning to recognise the familiar, to better structure new content and to learn more easily (Beyer 2011).

Geographical topics, concepts and phenomena; topographical information, regional geographical cases (places); or general geographical pieces of information, can be termed as the subject's *vocabulary*—partly as classic *subject matter*, and partly as *contextual knowledge* (Lambert 2013, p. 175). Key concepts, as conceptual knowledge, can be interpreted as the *grammar* of the subject. Thus, starting from single words of vocabulary, they can form a meaningful (systematic) overall structure of knowledge (Brooks 2013; Lambert 2013; Jackson 2006). Both vocabulary and grammar are necessary for forming geographical understanding.

With the key concepts as a leitmotif [or a 'clasp' that holds together the individual contents, (Fischer et al. 2007)] for recurring regularities in the problems dealt with in geography lessons, the cumulative build-up of subject knowledge and the acquisition of geographical thinking are supported—as are the accentuation, choice and structure of topics in teachers' lesson planning (Uphues 2013; Taylor 2011). In terms of a pair of geographical spectacles (or 'second order observation'), the key concepts can help direct the view of teachers and students to the central phenomena of the subject and their inner relationships (Demuth et al. 2005; Rhode-Jüchtern 2009). The different geographical concepts each provide their own spectacles which, when directed to the same topics and facts, identify their own details and create meaning (Fischer et al. 2007); in their entirety, they provide a geographical view.

These perspectives particularly help in grasping, classifying and establishing geographical understanding of new facts. In addition, the key concepts are taken as a basis for lesson planning and, supported by a step-by-step build up of the understanding of the concepts, make things transparent for the learner (Demuth et al. 2005). Starting with fundamental commonalities in different topics, through ongoing differentiation, an increasingly abstract understanding and insight into the transfer of knowledge to other contexts can be achieved (Beyer 2011), thus counteracting the critique of additive learning in geography lessons. By demonstrating consistent structures in the phenomena dealt with, a contribution to interconnectedness and powerful thinking can be made that helps to formulate a geographical enquiry of the world (Taylor 2011). Students are encouraged to apply their knowledge in various situations and for the acquisition of new understanding (Bünder et al. 2003). Key concepts order, interconnect and explain the topics, models and theories of geography [concepts as *classifiers* (Taylor 2008)], not only from a functional perspective. They also help in planning the lesson and aid learners in developing a deeper understanding of the subject (Brooks 2013).

5.3 Overview of Key Concepts and the Need for Further Development

As in the English-language context [see overview in Taylor (2008)], there is a wide range of key concepts for geography discussed in German discourse. In view of the historical development and paradigmatic diversity, as well as the epistemological span between physical and anthropogenic geography, the difficulty of formulating universally valid key concepts is understandable. For this reason, only selected concepts will be discussed in the following sections. These can only be understood as interpretations of the subject core that are currently valid and significant. Some central continuities that are common to the majority of ties to the subject are briefly outlined.

5.3.1 Human-Environment Systems

The *human environment* couplet is of prime significance in geography teaching today. Spatial observation and analysis of space in the human-environment system aims to examine the 'interactions between the social space and the natural space subsystems' (Rempfler and Uphues 2011a, p. 22). In the integrated processing of topics in the transition between natural and social scientific domains, a unique feature of geography is seen (Rhode-Jüchtern 2009). In contemporary geography the temporal, spatial and social interdependency of challenging problematic situations in the 21st century are effectively explained at the large scale and not by simple, linear causal relationships.

The thinking in (and about) the systems required to solve geographical problems is based on various conceptions about what a system is understood to be. A classic definition, primarily for physical geography, is still valid today even though it dates back to the work of Bertalanffy in the 1960s (Rempfler and Uphues 2011b). Here systems are 'understood as the totality of elements, characteristics or parts that are related to each other and interact with each other, so that they can be delineated from an environment' (Egner 2011), thus building a functional unit (Rempfler and Uphues 2011a). The inner relationship between the elements, their interaction, and the spatial arrangement of the elements to each other, constitutes a system's structure (Hard 1973). Grasping this structure, and the impact of such interrelationships, is a key proficiency interlinked in systems operational thinking (Köck 2001). At the same time, there are very different views in geography about the nature of systems—drawing partly on Luhmann's system theory as societal theory, which cannot be delved into here due to lack of space (Rempfler and Uphues 2011b). The various system concepts point out the difficulty of formulating cognitive skills for dealing with systems when various (partially hidden) ideas are present in the matter. It should be noted that the human-environment systems observed display a high degree of complexity, be it through the number of elements to be observed or the assumptions about the nature of a system as an open, dynamic, autopoietic and emerging construct (Rempfler and Uphues 2011b).

Geography now formulates its own claim to overcome this given complexity of systematic interrelationships (Rhode-Jüchtern 2009). For that, an interlinked or interlinking form of thinking is required (Köck 2001). Such system thinking can be understood to be the ability 'to describe, reconstruct and model a complex area of reality as a system and based on the modelling, provide explanations, come up with prognoses, and design, assess and implement possible actions' (Rempfler and Uphues 2011a). As skills which make up this ability, the dimensions *system*

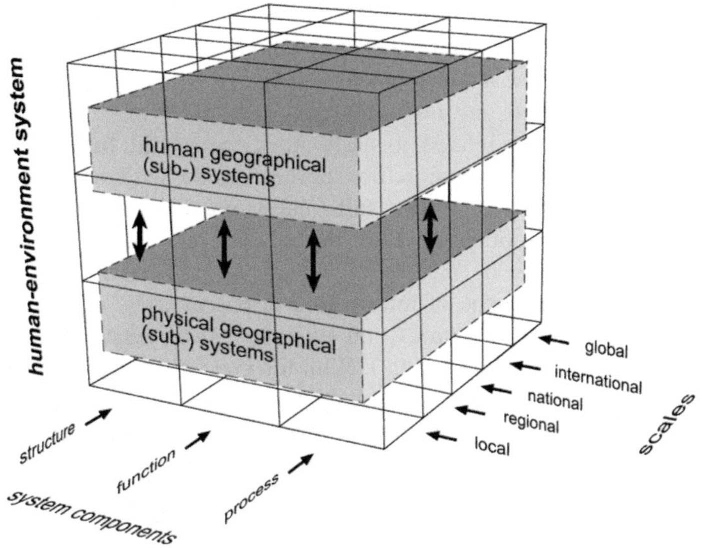

Fig. 5.1 Key concepts in German educational standards for geography. *Source* DGfG (2012, p. 11)

organisation, system behaviour and *system-appropriate action intention* are shown in a competence model (Mehren et al. 2014). The aptitude for these intellectual operations correspondingly represents an integral part of the German educational standards in the subject of geography (DGfG 2012).

The *cube model*, which describes the human-environment system, is depicted in Fig. 5.1. It substantiates the key concept *system* formulated in the German educational standards. In addition to the compartmentalisation in *human and physical geographical subsystems* and the presentation of their interactions as a central analysis approach, an axis is labelled with the *scale level*—from local to global (DGfG 2012). The respective systems, or subsystems, their interactions and individual system components can be examined on these scale levels. The system components *structure, function and process* as key concepts for the understanding of systematic relationships will be briefly explained.

5.3.2 Structure|Function|Process

From the spatial order of the system elements arises the **structure** of the systems, or subsystems (DGfG 2012; Borsdorf 2007). Phenomena that are physical-material (e.g. urban development features, soil quality or the respective infrastructure), or that can be experienced mentally or subjectively (e.g. industrial and cultural evidence, graffiti, standards in public space), are both conceivable as structural elements of the system. Their apportionment and spatial pattern give the structure of the system observed (Bette and Fögele 2015).

Since the elements are related to each other, they fulfil **functions** for other elements 'e.g. climate for vegetation, transport routes for settlements, the relief for transport routes' (DGfG 2012). The soil quality previously mentioned fulfils a function for agriculture, both are thus related to each other. The relationship between the elements can be the result of material connections, or can be substantiated through an exchange of information or energy.

Processes underlie the structures and functions; their genesis is attributed to spatial impact changes (Borsdorf 2007). With the system's dynamics, the elements as well as their relationships (hence, functions) go through processes—and with them, the system in general. For example, soil quality can change through intensive farming, the process of soil compaction causes changes in the soil structure with conceivable consequences for agricultural use. This shows that in the spatial examination of structures, functions and processes, the interplay among the concepts is also always to be considered. From this perspective, a systematic analysis of the human environment system can evolve (Bette and Fögele 2015).

Results of empirical studies (for a brief overview, see Fögele and Mehren 2015) show that for geography teachers, the presentation of human-environment systems and their significance is likewise central to geographical analysis. A survey of the 42 participating teachers carried out at the beginning of the aforementioned training courses also confirm this finding. The examination of spatial structures, their relationship to each other and the focus on space-defining processes all play a role for the teachers. All the more significant is the finding that this does not apply for the key concept of the four geographical concepts of space/place detailed below, equally important for the principles of teaching geography applied in Germany. The four concepts can be divided into two physical and mental concepts each. A significant finding is that the mental models play no role in the teachers' presentations, attention is given only to the physical-manifested space (see Fögele and Mehren 2015). This insight is the key motivation for emphasising the function of the hitherto little regarded mental-space concepts.

5.3.3 Four Concepts of Space and Place

There is extensive work available on the meaning of the two concepts space and place in the English-language context, which cannot be discussed here. However, there are a number of examples available in the German context on the study of space (Rhode-Jüchtern 2009). The systematisation, which is taken up below and which presents important groundwork for the reception of the space concepts in the principles of teaching geography, is attributed to Wardenga (2002). Wardenga has primarily contributed to the presentation of the epistemological and ontological development in the domain of geography as a subject. This chapter takes up the subsequent reflections on the utilisation of the concepts for the geographical analysis of space in lessons (Hoffmann 2009) and later portrays their added value for geographical thinking by means of a concrete example.

Space is considered as a largely accepted (and constitutive) category of geographical knowledge (Borsdorf 2007). 'Spatial order' (Rhode-Jüchtern 2013) is positioned as a basic theory of the subject and 'looking at the world spatially' (de Blij 2012) is also formulated as an original geographical perspective. Within the science of geographical space, the concepts in this category have regularly changed in the course of the subject's development—which can be read as an expression of its paradigmatic development. The space concepts are said to be complementary to one another, each focussing on different phenomena or following their own scientific theoretical assumptions (Hoffmann 2009).

Basically, four ideas about space are distinguished: space as a container, space as a system of relationships, space as a category of perception, and space as a construct (Wardenga 2002; Curriculum 2000+). This differentiation is also included in the educational standards (DGfG 2012).

If spaces are regarded as tangible, existing entities, this correlates to the metaphor of (space as) a *container*. Factors are examined for their interdependency, bundled under the concept of landscape as a space segment, and described and explained as a unique, tangible whole under the assumption that objective knowledge is possible (Wardenga 2002; Egner 2010). A likewise positivist interpretation underlies research into our understanding of space as *systems of relationships*. Here, too, the focus lies on physical-material elements, whereby spatial relationships, distances and locations are explored. In a nomothetic approach, functional connections between space elements become visible, making prognoses possible (Wardenga 2002).

Following behavioural science and constructivist approaches, an orientation to the subject takes place under the concept of space/place as a *category of perception*. Spaces, like societies and realities, are to be considered as potentially pluralised and perception-dependent dimensions (ibid.). Linked to this is the perspective of spaces as *artefacts of social construction acts* (ibid.). If one understands spaces to be socially construed entities, firstly, the subject is now the active doer and secondly, spaces (or places) are produced or constituted by this action and in the communication between subjects (e.g. collective action in society) (Wardenga 2002).

Geographical concepts of space/place provide a tool that structures four geographical views of a chosen space as well as problem areas, whereby each perspective has its own type of questioning and addresses specific categories of issues (Rhode-Jüchtern 2009). They are in agreement with the concepts of the cube model (see Fig. 5.1), which means they are not competing but rather are seen as a complementary analytical grid. Table 5.1 suggests exemplar questions that could be associated with the four concepts.

Table 5.1 Geographical concepts of space/place with exemplar questions

	Space concept	Exemplary questions (here in the context of natural hazards)
Physical material, measurable space Object-orientation and order of things	Space as container	• How do (which) geofactors work in causing a landslide? • What are the characteristic features of the space (e.g. relief, climate, population structure)? • What space changes can be expected from human effects?
	Space as a system of relationships	• What are the relationships between the factors in the space at the various levels and how did their interaction contribute to the creation of the landslide? • How do local, regional and global conditions relate to each other (e.g. local ownership situation, regional supply, national building regulations), are there any dependencies? Are conditions between different spaces comparable? • How is the endangered area included in the regional and national infrastructure? What exchange relationships exist?
Mental space/place (perception and communication) Subject-orientation and order of views	Space/place as a category of perception	• How was the landslide perceived and by whom? How was it assessed? Are there different assessments? • How do the assessments differ per group (e.g. age, profession such as politician, merchant or farmer, victims…) and how do they differ (e.g. subsequent decisions)? • What is the relationship between the perceived and measureable aspects of the space/place?
	Space/place as construct	• How does a landslide become a natural catastrophe? • Who is involved? Who is a victim? What are the consequences for dealing with the event? • By whom and how will the landslide be reported? What influence does that have on the perception of the event?

Source Author, based on Hoffmann (2009), Rhode-Jüchtern (2009)

5.4 Reflections on a Cognitively Activating 'Culture of Tasks'

The design of a lesson that is cognitively activating, and encourages independent problem solving in real-life contexts, allows the application of prior knowledge which benefits progress in learning. Thinking is aided in lessons with the support of activating tasks (Hieber et al. 2011). The preceding features are shown accordingly as the cornerstone of a new 'culture of tasks'. In addition, other dimensions, which cannot be discussed here in detail, determine the quality of the exercise—such as the clarity of its objective, its potential for metacognitive learning phases (reflection on the elaboration), or the degree of its place in the students' lifeworld

(Hieber et al. 2011). Within the framework of this chapter, a central feature is the orientation to geographical key concepts and how these can be helpful in activating exercises for structuring work that is mostly independent.

This way of learning overlaps with *geographical enquiry* (Mitchell 2013; Taylor 2008). Four features serve to characterise this approach: 'starting by creating a need to know, then using data, making sense of the data and finally reflecting on learning' (Ferretti 2013). These steps are initiated through activating tasks, each providing possibilities for the integration of key concepts.

Curiosity about worthwhile, lifeworld problems should be aroused; the intrinsic interest of wanting to work out the exercise (Taylor 2008) and take the opportunity given to access prior knowledge such that hypotheses on complex and possibly controversial issues can be formulated (Ferretti 2013). Key concepts help learners pose geographical enquiries, identify key issues, encourage assumptions and thus provide an orientation aid within the context of 'challenging enquiries' (Jackson 2006). As a result, available material is processed and further material can be collected and sorted by functional aspects and relevance for the problem-solving process (Ferretti 2013). Subsequently, relationships are worked out in the form of structural diagrams (concept maps), with selected key concepts as hubs. Via several metacognitive processes (such as interpretation, comparison and contrast, the establishment of relationships, seeking functional order, and describing or synthesising supported by key concepts) meaning is generated such that decisions and answers to hypotheses formulated at the outset are found (Mitchell 2013; Ferretti 2013). After the subsequent critical reflection (on the sources, one's own approach and the guiding categories and concepts, on what was learned and its value, and on the ability to transfer this to other issues), this knowledge of the *enquiry-approach* in terms of a circular process is incorporated in the processing of future cases (Ferretti 2013). In this way conceptual knowledge should contribute to the promotion of geographical skills and methods as well as encouraging metacognitive strategies that allow a transfer to other issues (Laske and Schuler 2012).

5.5 Practical Implementation of the Four Concepts of Space/Place: The Climate in London (Two Examples)

5.5.1 Exercise Without Geographical Concepts of Space/Place

The following activity is established:

'*Please analyse the climate chart of London and compare it to the chart of Rome.*'

For this exercise, the students have a climate chart of London and one of Rome. These need to be read correctly since the interpretation of the charts permits a comparison of the climate conditions of the two cities and possibly a transfer to the corresponding regions in Europe. This is a typical exercise in textbooks for geography lessons in Germany, the practice of dealing with climate charts can be seen as a classic, methodical form of instruction in geography lessons. But students are not cognitively challenged, so the knowledge and skill acquired remain at a low level of schematic application. It is expected that the ability to interpret a climate chart be acquired in the geography classroom—but it should not remain an end in itself. Furthermore, the exercise posed contains no practical application to life; it is narrow and offers no opportunity for the students to progress or develop their own creative thinking (Leisen 2006).

The following example is to show how, by using a different version of the exercise, a challenging and worthwhile enquiry can be developed. With the aid of key concepts, more powerful thinking can be acquired, so that in terms of geographical enquiry instead of exercise completion, this can be experienced as 'a journey towards solving a puzzle or constructing an answer' (Taylor 2008, 50).

5.5.2 Exercise with Geographical Concepts of Space/Place

Here the activity is:

'*Why do we think that it is always raining in London*?'

This exercise is carried out based on the students' assumptions that are formulated in the working material step-by-step along the four concepts of space/place. Materials such as climate charts, media depictions of London, further explanations about climate conditions in Europe, or a map of Europe for climate classification should be made available for selection by the students in small groups. The starting point is the question about the background of a realistic idea about climate conditions in London.

1. Assumption: Even without supporting materials, it is assumed that it rains a lot in London. To prove or disprove this hypothesis, information derived from the climate chart is also required. This information presents a feature of space as a container and deals with the geofactor 'climate'. An annual accumulation of rainfall of somewhat more than 600 mm does not, however, answer the initial question, so further information needs to be gathered.
2. Assumption: Only by comparing with other places can it be construed how to assess an annual accumulation of 600 mm. Comparisons, relationships and conditions between physical spaces are the subject of the second space concept: space as a system of (spatial) relationships. At this point, the interpretation of the climate chart for Rome shows that in total, and on average, there is more than 850 mm of precipitation here annually. This insight is surprising and is contrary to previous assumptions.
3. Assumption: Based on one's own prior knowledge about London and initiated by other material, the third hypothesis is posed that we believe, for example, that it is always raining in London because it is frequently depicted in the media as

being a rainy city. But even this assumption on the basis of space/place as a construct does not go far enough. It does not explain why London is often portrayed in this way.

4. Assumption: London is perceived to be a wet city. But why? Information about space can be related to people's perceptions. Rome is a travel destination particularly in the warmer months when rain hardly plays a role here. This may be a first, temporarily satisfactory, answer to the initial question. At the same time, the answer is not satisfactory since it is based solely on the comparison between the two spaces/places—but even without the comparison to the conditions in Rome, the perception of London as a rainy city appears to have been significant from early on.
5. Assumption: It is not the absolute rainfall that is the key, but the way in which it falls! The precipitation regularly falls as drizzle and in conjunction with relatively constant precipitation all year, this leads to less frequent periods of dry conditions. This finding explains, to a large extent, the fourth assumption about London being a wet city. This in turn explains the third assumption, and why the city is portrayed in this manner in German media and culture. (It is worth noting that English people do not necessarily perceive of London in this way—hence their saying 'It always rains in Manchester').

In the end, this study has stimulated reflection on the climate conditions in London and explained how stereotypes originate. In Fig. 5.2 this cycle of awareness is shown schematically.

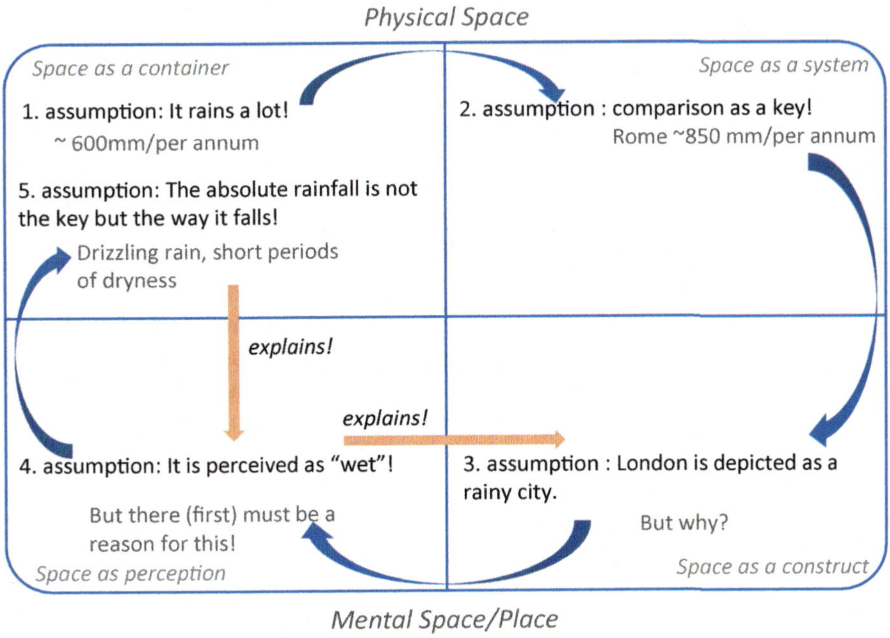

Fig. 5.2 Possible assumptions using the four space concepts. *Source* Author's illustration

As this example shows, even in a lesson that focusses on key concepts, content concepts (climate zones, precipitation regimes, etc.) and skills (map reading, interpretation of pictorial media, interpretation of climate charts, etc.) remain important (Taylor 2008). However, they have been advanced to a further level of knowledge. A frequently espoused typology of various forms of knowledge distinguishes factual, conceptual, procedural and metacognitive knowledge (Krathwohl 2002): 'We have reserved the term Factual Knowledge for the knowledge of discrete, isolated 'bits of information' and the term Conceptual Knowledge for more complex, organized knowledge forms' (Anderson et al. 2001). By integrating the key concepts with activating exercises or tasks, a constructive use of factual and procedural knowledge now takes place.

The resolution of the second exercise indicates how the four concepts of space/place as geographical key concepts in the German-speaking context help, from different perspectives, to formulate questions about the subject, establish connections between these dimensions and thereby allow a largely independent development of a worthwhile enquiry. In the outline of the examination of this learning exercise, the phases of the above-mentioned geographical enquiry (Ferretti 2013) are clearly identifiable. This approach, firstly, is more challenging and secondly, goes into more depth than the process previously outlined.

5.6 Outlook

Key concepts pertain to several areas of a teacher's professional skills. According to Shulman (1986), they can be classified as subject-specific content knowledge and as pedagogical content knowledge. While, as has been shown, epistemological and ontological positions on the subject and subjective theories on teaching and learning also have bearing on geography education, it is clear that their implementation in geography lessons is challenging. Firstly, there are other aspects of lesson planning to be included, such as the exercises for students to complete, and secondly the further development of knowledge and teachers' beliefs cannot be achieved through short-term teacher training courses. For this reason longer periods of professional development for teachers—in the form of sustained training programmes with phases for input, assimilation, practical testing and development of teaching practice, should be initiated. A list of ten criteria for the design and implementation of more effective teacher training, derived from empirical findings, builds an important groundwork (Höhnle et al. 2016; Fögele and Mehren 2015). The exemplar exercises previously presented emerged from a training course through collaboration with participating teachers. In this context, additional intensive reflective discussions with the teachers were encouraged. Figure 5.3 contains a sample excerpt from them.

For some participants, the examination of geographical key concepts has led to new reflections about the nature of their subject and on the objectives of their own teaching. These experiences indicate that key concepts can be used as an

Bf: *We used to design and organise our lessons from the content. In my case the development, the enhancement of the pupils' knowledge- that is what I have in mind. I consider what we can learn from this topic, now I have an eye for the big picture!*
Em: *└ Before I am able to do that- I mean it is a long process!*
Bf: *└ Yes! And it is difficult. But we should keep in mind what they really should - ahm- should be able to understand!*
Em: *But we did this before! Didn't you include considerations about the frame of the content?*
Bf: *Not with this complexity, I never pondered on geography in that way!*

Fig. 5.3 Excerpt from a reflective discussion with teachers. *Source* Author's survey

opportunity, and as a tool for powerful thinking to grasp the 'big picture' of more complex geographical phenomena. The finding that geography teachers can experience 'overload' in their efforts to address their subject's complexity, particularly when teaching topics at the global scale (Hof and Hennemann 2013), suggests that geographical key concepts can help deal with complex issues.

References

Anderson, L., Krathwohl, D., Airasian, P., Cruikshank, K., Mayer, R., & Pintrich, P. (2001). A taxonomy for learning, teaching and assessing: A revision of Bloom's taxonomy of education objectives. New York: UA.

Bennett, T. (2005). Progression in geographical understanding. *International Research in Geographical and Environmental Education, 14*(2), 112–132.

Bette, J., & Fögele, J. (2015). Mit Basiskonzepten Aufgaben strukturieren und fachliches Denken diagnostizieren. *Am Beispiel von Islands Energieerzeugung, Praxis Geographie, 45*(7), 34–39.

Beyer, I. (2011). Natura—Biologie für Gymnasien: Basiskonzepte, Sekundarstufe I und II (1. Aufl., [Dr.] 1). Stuttgart, Leipzig: Klett.

Borsdorf, A. (2007). Geographisch denken und wissenschaftlich arbeiten (2. Aufl). Berlin: UA, Springer, Spektrum Akad.-Verl.

Brooks, C. (2006). Geographical knowledge and teaching geography. *International Research in Geographical and Environmental Education, 15*(4), 353–369.

Brooks, C. (2013). How do we understand conceptual development in school geography? In D. Lambert & M. Jones (Eds.), *Debates in geography education* (pp. 75–88). Abingdon: Routledge.

Bünder, W., Demuth, R., & Parchmann, I. (2003). Basiskonzepte: Welche chemischen Konzepte sollen Schüler kennen und nutzen? *Pdn-ChiS, 52*(1), 2–7.

Curriculum 2000+ and Deutsche Gesellschaft für Geographie, DGfG (2002) Grundsätze und Empfehlungen für die Lehrplanarbeit im Schulfach Geographie: Arbeitsgruppe Curriculum 2000+ der Deutschen Gesellschaft für Geographie.

de Blij, H. (2012). *Why geography matters: More than ever* (2nd ed.). Oxford: Oxford University Press.

Demuth, R., Ralle, B., & Parchmann, I. (2005). Basiskonzepte—eine Herausforderung an den Chemieunterricht. *CHEMKON, 12*(2), 55–60.

DGfG, German Geographical Society. (2012). Educational standards in geography for the intermediate school certificate. Bonn: Selbstverlag Deutsche Gesellschaft für Geographie.
Egner, H. (2010). *Theoretische Geographie (1. Aufl)*. Geowissen kompakt, Darmstadt: WBG.
Egner, H. (2011). Systemtheorie und Geographie. In H. Gebhardt, R. Glaser, U. Radtke & P. Reuber (Eds.), *Geographie. Physische Geographie und Humangeographie* (2nd ed., p. 1088). Heidelberg: Spektrum, Akad. Verl.
Ferretti, J. (2013). What happened to the enquiry approach in geography? In D. Lambert & M. Jones (Eds.), *Debates in geography education* (pp. 103–115). Abingdon: Routledge.
Fischer, H., Glemnitz, I., Kauertz, A., & Sumfleth, E. (2007). Auf Wissen aufbauen: Kumulatives Lernen in Chemie und Physik. In E. Kircher, R. Girwidz & P. Häussler (Eds.), *Springer-Lehrbuch. Physikdidaktik* (pp. 657–678). Theorie und Praxis, Berlin, Heidelberg: Springer-Verlag.
Fögele, J., & Mehren, R. (2015). Implementing geographical key concepts. Design of a symbiotic teacher training course based on empirical and theoretical evidence. *Review of International Geographical Education Online, 5*(1), 56–76.
Hard G (1973) *Die Geographie: Eine wissenschaftstheoretische Einführung. Sammlung Göschen, 9001*. Berlin: W. de Gruyter.
Hieber, U., Lenz, T., & Stengelin, M. (2011). (Sich) geographische Aufgaben stellen: Neue Aufgabenkultur im kompetenzorientierten Geographieunterricht. In T. Lenz (Ed.), *Aufgaben stellen- Kompetenzen fördern, Geographie heute, Heft 291* (pp. 2–9).
Höhnle, S., Fögele, J., Mehren, R., & Schubert, J. (2016). GIS teacher training- empirically based indicators of effectiveness. *Journal of Geography, 115*(1), 12–23.
Hof, S., & Hennemann, S. (2013). Geographielehrerinnen und -lehrer im Spannungsfeld zwischen erworbenen und geforderten Kompetenzen. *Eine empirische Studie zur zweiphasigen Lehramtsausbildung, Zeitschrift für Geographiedidaktik/Journal of Geography Education, 2*, 57–80.
Hoffmann, K. (2009). Mit den nationalen Bildungsstandards Geographieunterricht planen und auswerten. *Zeitschrift für Geographiedidaktik/Journal of Geography Education, 3*, 105–119.
Jackson, P. (2006). Thinking geographically. *Geography, 91*(3), 199–204.
Köck, H. (2001). Typen vernetzenden Denkens im Geographieunterricht. *Geographie and Schule, 132*, 9–15.
Krathwohl, D. (2002). A Revision of Bloom's taxonomy: An overview. *Theory into practice, 41* (4), 212–260.
Lambert, D. (2013). Geographical concepts. In M. Rolfes & A. Uhlenwinkel (Eds.), *Didaktische Impulse. Metzler Handbuch 2.0 Geographieunterricht, Ein Leitfaden für Praxis und Ausbildung* (pp. 174–181). Braunschweig: Westermann.
Laske, J., & Schuler, S. (2012). Mit Geographie denken und Probleme bearbeiten kernen: Aufgaben im problemlösenden Geographieunterricht. In Westermann Schroedel Diesterweg (Ed.), *Wie wir lernen, Aufgabenkultur im Geographieunterricht* (pp. 12–17). Praxis Geographie Jg. 42, 2012, Heft 12, Braunschweig.
Leisen, J. (2006). Aufgabenkultur im mathematisch-naturwissenschaftlichen Unterricht. *MNU, 5*, 260–266.
Mehren, R., Rempfler, A., & Ulrich-Riedhammer, E. (2014). Denken in komplexen Zusammenhängen: Systemkompetenz als Schlüssel zur Steigerung der Eigenkomplexität von Schülern. *Praxis Geographie, 44*(4), 4–8.
Mitchell, D. (2013). How do we deal with controversial issues in a 'relevant' school geography? In D. Lambert & M. Jones (Eds.), *Debates in geography education* (pp. 232–243). Abingdon: Routledge.
Morgan, J. (2012). *Teaching secondary geography as if the planet matters*. Abingdon: Routledge.
Rempfler, A., & Uphues, R. (2011a). Systemkompetenz und ihre Förderung im Geographieunterricht. *Geographie and Schule, 33*(189), 22–26.
Rempfler, A., & Uphues, R. (2011b). Für ein adäquates Verständnis von Geosystemen. *Geographie and Schule, 33*(189), 4–10.

Rhode-Jüchtern, T. (2009). Eckpunkte einer modernen Geographiedidaktik (1. Aufl) Seelze, Kallmeyer.
Rhode-Jüchtern, T. (2013). Geographieunterricht: Weltverstehen in Komplexität und Unbestimmtheit. In D. Kanwischer (Ed.), *Studienbücher der Geographie, Geographiedidaktik, Ein Arbeitsbuch zur Gestaltung des Geographieunterrichts; mit 4 Tabellen* (pp. 21–33). Stuttgart: Borntraeger.
Shulman, L. (1986). Those who understand: Knowledge groth in teaching. *Educational Researcher, 15*(2), 4–14.
Taylor, L. (2008). Key concepts and medium term planning. *Teaching Geography, 33*(2), 50–54.
Taylor, L. (2011). Basiskonzepte im Geographieunterricht. Schlüssel, um die Welt besser zu verstehen und den Unterricht besser zu planen. *Praxis Geographie, 7*, 8–15.
Uphues, R. (2013). Basiskonzepte. In G. Obermaier & D. Böhn (Eds.), *Didaktische Impulse* (pp. 22–23). Begriffe von A-Z, Braunschweig: Wörterbuch Geographiedidaktik.
Walshe, N. (2007). Understanding teachers' conceptualisations of geography. *International Researc in Geographical and Environmental Education, 16*(2), 97–119.
Wardenga, U. (2002). Räume der Geographie: Zu Raumbegriffen im Geographieunterricht. *geographie heute, 23*(200), 8–11.

Chapter 6
What Type of Geography Do We Teach? from Theoretical-Conceptual Weaknesses to Underestimation of Spatial Experience. Chilean Teachers' Views on Teaching Geography

Andoni Arenas-Martija, Patrico Peréz-Gallardo, Victor Salinas-Silva and María José Otero-Auristondo

6.1 Introduction

Over the last decade, several countries have tried different processes in geography education to develop geographical thinking in schools. These processes have generated many discussions regarding the importance of geography (Gómez, et al. 1988; Republic of Botswana 2000). These discussions provide both challenges and opportunities to further promote and develop geographical thinking (Bahbahani et al. 2010; Biddulph et al. 2015; The Critical Thinking Consourtium 2010–2011; Lidstone and Williams 2006; MINEDUC 1999, 2012, 2013). This chapter contributes further to these discussions through its focus on a research project: *What geography do we teach? The meaning of the Geographical phenomena in instructional practices of Teachers in the Teacher of Teachers network and Rural Micro Centre*[1] which was conducted in 2013–2014 with geography teachers in Chile.

6.2 Research Approaches

The main research question addressed in the project originates from our consideration of four main elements of the relationship between education and geography. These are:

[1]Research Project Number 11130400-2013, National Commission on Science and Technology, (CONICYT). Ministry of Education, Chile.

A. Arenas-Martija (✉) · P. Peréz-Gallardo · V. Salinas-Silva · M.J. Otero-Auristondo
Instituto de Geografía, Pontificia Universidad Católica de Valparaíso, Valparaíso, Chile
e-mail: andoni.arenas@pucv.cl

C. Brooks et al. (eds.), *The Power of Geographical Thinking*,
International Perspectives on Geographical Education,
DOI 10.1007/978-3-319-49986-4_6

- The development of geographical science
- Geographical knowledge as curriculum content
- Identifying what happens in the geography classroom
- Understanding the significance of what happens in the geography classroom.

With regards to the development of geographical science we argue that there has been a paradigmatic shift in contemporary geography which has led to humans being its primary focus. We believe that there are a series of identifiable meanings about "the geographical" both in academic and school geography and amongst individuals and societies in terms of how they inhabit space.

With regards to geographical knowledge as curriculum content, we argue that geography has the capacity to generate knowledge that is useful to society and directly contributes to the scientific literacy of a country's inhabitants. We believe that people who are able to understand natural and social phenomena from a geographical perspective assist in improving understanding of the problems affecting the space in which we live.

With regards to the significance of identifying what happens in the geography classroom in Chile, we argue that research has not analysed what happens in sufficient depth. As a result there is little pedagogical knowledge for teachers based on research collected in classrooms.

With regards to understanding the significance of what happens in geography classrooms, we argue strongly that high quality geography teaching and the promotion of geographical thinking depend on the connections that are made between students' and teachers' understanding of the curriculum. In other words, the meanings they ascribe to the geographical are influenced by their personal geographical experiences.

Thus this research focuses on the broader phenomenon of what affects how Chilean geography teachers teach their subject. The research therefore considers the complex web of cultural meanings within educational and social practices and the important role played by teachers and students in these practices. This requires deeper understanding of the web of meanings associated with the geographical and current teaching practices and learning, given that this vast web is configured and shaped by everyday experiences of geographical space. The research question therefore focuses on teachers who teach a curriculum that they *know* and also *experience* (Tuan 2007; Garrido 2009) and who therefore ascribe meanings to the geographical content they teach and to how students learn.

The research examines specifically the practices of teachers belonging to the Teacher of Teachers Network (TTN) and the Rural Micro Centres Network (RMC). TTN is a programme operated by Chile's Ministry of Education aimed at strengthening the teaching profession by sharing the skills of teachers who have been accredited as teachers of excellence. These teachers contribute to the professional development of their colleagues and the schools in which they work.

RMC was introduced in 1992 as a strategy to improve pedagogical practices in the classroom by bringing together teachers from rural schools through a series of professional development workshops with support from supervisors from the Ministry of Education.

6.3 Research and Theoretical Framework

The research question posed was approached from the following theoretical perspectives:

Teachers' pedagogical methods as enablers of student learning: Teachers participate in knowledge exchange and construction of meaning in the classroom. Yet in order to teach a specific area of knowledge, the teacher has to build bridges between the meanings contained in the curriculum, provided in the classroom, and those that students already have (Shulman 1993), referring to both formal and experience-based meanings that shape the way we think.

Life experience as a source of geographical knowledge and thinking: Geographical knowledge and the geographical thinking it sustains are more than merely objective and measurable knowledge; they are also a dimension of living in the world and a way of understanding it, filling the objective space with numerous experiences and expectations where meanings multiply, occur and counter one another (Nogué 2007).

Geography as an educational programme: As a result of developments within the academic discipline, geography is now widely recognised as a valuable subject in schools (Reinfried et al. 2007) that encompasses a multitude of issues and theoretical-methodological possibilities (Palacio 2011). Geography education has benefitted from a renovation of disciplinary knowledge and new pedagogical perspectives focused on achieving relevant learning. This has coincided with a pedagogical shift aimed at clarifying the epistemological dimension of teaching and learning processes (Prats 2003) and rearticulating theory and practice by focusing on the way in which knowledge is taught and learned (González 2002), broadening understanding to include what happens outside the classroom while remaining centred around what occurs in the classroom.

Geographical knowledge has been based on theoretical and practical constructs originating in different currents of thought: The contexts and temporality of these currents have operated on the basis of paradigmatic agreements, which for geography has meant reformulating the object of study and methodological approaches on more than one occasion (Bosque and Ortega 1995). The different *moments* of geographical knowledge range from pre-scientific geography closely associated with the discovery of new territories to scientific geography that sought legitimacy as a science (traditional geography) through a very systematic, pragmatic and exact approach. This included geographical theories such as determinism, possibilism, regionalism and neo-positivism (Moraes 2006). With the dissemination of *critical theories* in recent decades, the science of geography has taken on the study of geographic space that is more closely linked to societal conditions, addressing questions related to spatial dynamics.

In Chile, the educational curriculum does not reflect the evolution of geography as a discipline: The existing geography curriculum is more akin to an enumerative, descriptive, generalised geography focused on teaching formal knowledge. Students' informal experiences of geographical space (Garrido 2009), which reflect the everyday dynamics and problems of our society, are not part of the learning process. In this sense, school geography is primarily influenced by the legacies of traditional geography (Arenas and Salinas 2013).

6.4 Research Design

The general question that guided the study was: What is the understanding of members of the Teacher of Teachers Network (TTN) and the Rural Micro Centers Network (RMC) about what constitutes the geographical manifested in their teaching practices? Specifically:

- What are the different types of meaning of geographical topics that teachers have in the Teacher of Teachers Networks (TTN) and Rural Micro Centres (RMC)?
- How are these meanings used in teachers' pedagogy?
- What are the relationships that exist between the meanings of the geographical phenomena and teaching practices of teachers of TTN and RMC?

To answer these questions, an interpretative and exploratory study was designed to understand phenomena associated with geography teaching and learning in schools as a first approach to a subject never explored before in Chile. Framed within the social constructivism of reality (Berger and Luckmann 1996) and the geographical dimension (Nogué and Romero 2006; Nogué 2007), the study examined teachers' understanding of what constitutes geography using an inductive approach and interpretive analysis. It was assumed that teachers' knowledge of the scope of geography education (Lidstone and Williams 2006) and its relevance in teacher training (Alexandre 2009) is epistemologically valid (Fenstermacher 1994). This was examined from a qualitative perspective in which human phenomena (including educational and geographical phenomena) develop in specific social contexts via constructions that are the result of social-spatial negotiations (Schwandt 1998).

6.5 Research Methods

The principal research method used consisted of instrumental and collective case studies since the composition of the selection was set to answer the research questions (Stake 1998, 2000). Data was gathered from forty-three rural primary teachers (multi-grade, multi-level and multi-subject), grouped in a network (TTN) within six rural micro centres across seven different municipalities. Two of the teachers work in state-subsidised private schools. This network allows the teachers from different schools to work collaboratively in monthly workshops within a yearly plan, an exception in the working philosophy of teachers in Chile. In addition, the urban teachers, members of the RMC from five different municipalities, included one teacher from a publicly funded school. The RMC is composed of teachers with the best qualifications in the Chilean educational system, who have access to funds for innovation in the classroom and for collaboration with other teachers.

Data was collected through semi-structured interviews which provided the significance of geography and geography as stated by teachers; one class observation per teacher recording the educational expressions which can be the utterances of meanings; and one characterisation questionnaire recording background context. This chapter presents only the results from the semi-structured interviews and class observations. The data was recorded using field notes and audio recordings, which were transcribed verbatim. The data collection period was April 2014–June 2014. The first phase of analysis, which this chapter describes, was conducted between August and December 2014.

Data was analysed in terms of its content and triangulation, using structural-explanatory content analysis techniques (Flick 2004). This was based on pre-existing categories to guide the analysis from the theoretical benchmarks and emerging categories, to secure what emerged from the data (Miles and Huberman 1994; Giddens 1997). Three dimensions were analysed: Elaboration and Re-elaboration of Geographical Knowledge, Manifestations of Geographical Topics in Teaching, and Significance of what is Geographical (as such subject-object) and Geography (such as formal science). The analysis categories are shown in Table 6.1. As mentioned in the Introduction, the results discussed in this chapter relate directly to the third dimension.

For this in-depth analysis, triangulation (Lincoln and Guba 1985) that is both methodological and theoretical (McKernan 1999) was used. The data collection techniques and analysis procedures were approved by the ethics committees of the Pontificia Universidad Católica de Valparaíso and the National Science and Technology Commission of Chile. The following results were grouped in emergent

Table 6.1 Analysis categories: significance of what is geographical and geography

Meanings of what is geographical (subject-object) and geography (such as formal science)
Theories and moments of geographical thought These are moments in the past dominated by structures of philosophical thought that support paradigms in scientific thinking. These moments in time tend to influence paths and trends in the dominant epistemology of scientific development and cannot be separated from the epistemology itself. These moments serve as guides and enable recognition of features in structures of thought, redefining their methodology, practice and theory
Scientific traditions of geography This primarily refers to each actor's personal understanding of the way in which geographical study and thinking should be developed. The dominant traditions in the evolution of geography are nomothetic, chorographic, chorologic and ideographic
Epistemological paradigms of geography A set of similar theories within a main structure. These theories share a common organisational thought scheme, methodology, tradition and scientific practices. They represent the fundamental ideas of epistemological currents. Examples of epistemological paradigms are positivism, neopositivism and structuralism
Social value of geographical knowledge Societies in general and educational communities and their participants have a series of expectations about geography and geography teachers; these expectations determine the value that is given to this school subject
Methodological models Generally understood as the means used to achieve a result. It should be understood that in addition to being essentially scientific-methodological elements, these are also relevant to teachers' conceptualisation of reality, the way they realise the teaching-learning relationship or their ideas regarding the methodological functioning of geography itself: inductive, deductive, analytical, synthetic, comparative and dialectical
Branch themes of geography Throughout the evolution of geography, geographers have taken different thematic approaches to spatial reality. As the subject-object study of this reality has developed along different lines, tensions have arisen among geographers in regard to the validity of the knowledge developed
Categories of analysis and geographical concepts This refers to any lens used to analyse geographical reality. Each lens has its own conceptual weight and focuses on different aspects, defining the subject-object of geography in different yet complementary ways: geographical space, place, landscape, region, territory, environment and geosystem

Based on Lacoste (1981), Harvey (1983), Santos (1990), Bosque and Ortega (1995), Gurevich (2005), Moreira (2006), Moraes (2006), Tuan (2007), Nogué (2007), Gallestegui and Galaz (2009), Garrido (2009), Capel (2012), Arenas and Salinas (2013)

categories raised from the analysis of data with the pre-existing categories (Table 6.1). The established relationships between the theoretical framework and the pre-existing analysis categories are presented in Fig. 6.1, whilst the relationships established between existing analysis categories and emergent categories are presented in Fig. 6.2.

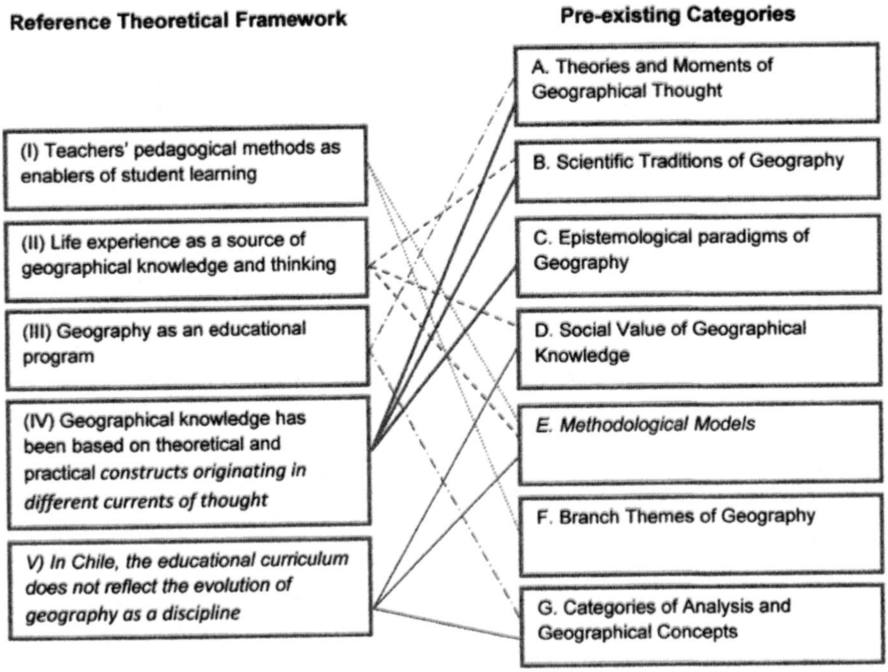

Fig. 6.1 Reference theoretical framework and pre-existing relationship categories (*Source* Prepared by the authors)

6.6 Research Findings

The main findings are grouped into categories and relevant subcategories.

6.6.1 Scientific Traditions of Geography and Methodological Models

Teachers emphasise geographical knowledge that focuses on spatial position/localisation. The importance given to spatial location is expressed through tasks that involve locating geographical spaces (local and global) such as regions, countries and continents, or locating features of these spaces using specific geographical coordinates. In this respect, preparation of certain learning objectives that reflect this focus were identified:

- Knowing where we are, our locality, our location
- Recognising that maps and plans are ways to represent places
- Locating places on the map, using geographical coordinates.

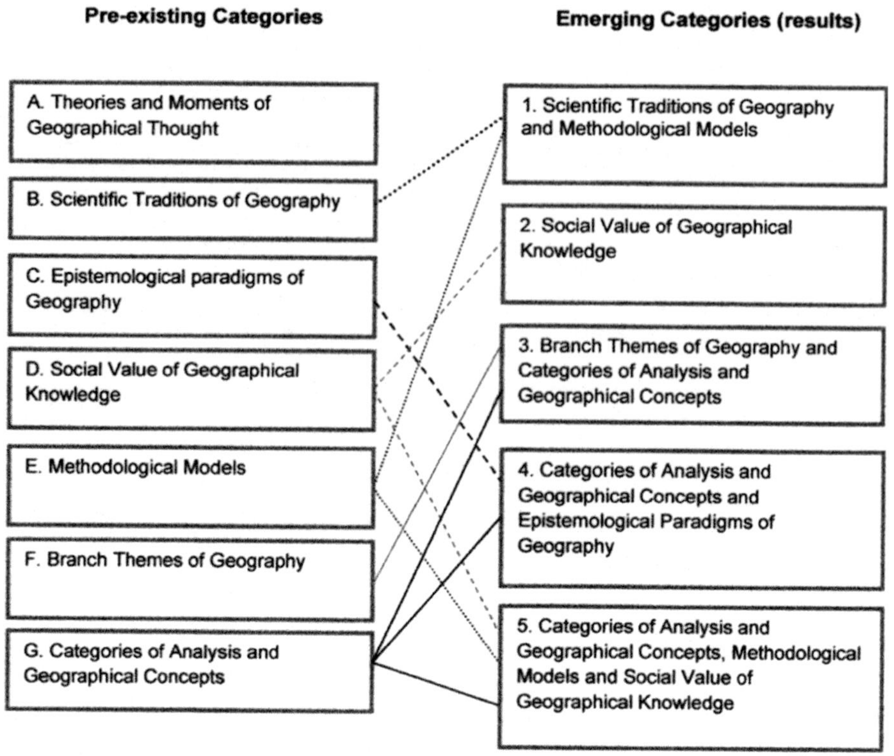

Fig. 6.2 Relationship between existing categories and emerging categories (results) (*Source* Prepared by the authors)

For example, one teacher who was teaching compass points asked her students: "Is Chile further north or south? Is it to the east or west?" (Class observations ID7). Another teacher began the lesson by introducing the concept of geography as follows:

> Geo means 'Earth' the teacher said, and then asked: "So what does geography mean?" Amongst the various responses given by students was the idea that it is the graphic representation of the earth. The teacher tells the students this is correct and that 'geography's key graphic tool is the map, which is viewed as a method of representation'.
>
> (Class Observations, ID7)

Another teacher began their lesson with a review of Chile's northern, southern, eastern and western boundaries, pointing out the neighbouring countries and the ocean (Class observations ID17). Relationships between localisation, scales of representation and understanding of geographical space were established by the teachers in a similar way. For example, one of the teachers taught spatial localisation by analysing it from the local to the universal level (locality-municipality-province-region-continent-planet). (Class observations ID23), using the resource

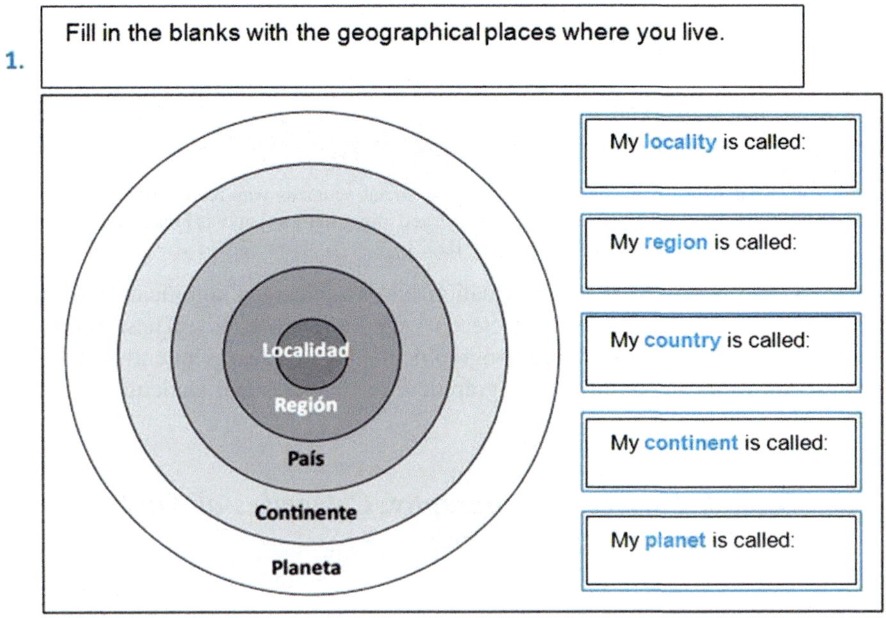

Fig. 6.3 Diagram of geographical places from the teaching module for multi grade rural schools (*Source* Education Ministry 2014 p. 3)

shown in Fig. 6.3. This resource is used to introduce the concept that a locality is contained (or located) in a larger geographical space called a region, which in turn is located in a larger space called a country and so on. It is used to teach students about their position/localisation within a self-contained geographical space that has absolute and fixed limits.

6.6.2 Social Value of Geographical Knowledge

In addition to considering position-localisation to be a central element in teaching geography, many of the teachers also emphasised its social value. In response to the question 'Why do you think it is important to teach geography?' one teacher replied: 'Geography is very important because the children will know where they are located, they will recognise different types of plants, places' (Interview ID11). According to another teacher, 'knowing your location, realising that certain details are important for knowing where you are in the world and also understanding that you are part of the world' (Interview ID33). Some of the teachers expressed the social value of geographical knowledge by recognising that their own geographical knowledge and understanding of how to teach geography is limited. They attributed this to their initial teacher training. One teacher commented:

> At university I was hardly taught any geography, it was always more world history, history of Chile. What I have learnt about geography are the things I have read, seen on the Internet: videos to watch in class and I have to check things first… (Interview ID11).

Several teachers expressed the opinion that the geography education they received in their initial training was limited and superficial. One teacher stated:

> For me, what I did in history, geography and social sciences was really basic, so what happens? The things I have learned, I have learned along the way, and as I encounter topics in the curriculum, I teach them *(Interview ID42).*

Others also remarked that this situation had not changed significantly and that currently schools required that they teach very little geography. These issues are clearly obstacles to understanding geographical reality and consequently hinder the construction of meanings of the geographical for teachers and students alike.

6.6.3 Branch Themes of Geography, Categories of Analysis and Geographical Concepts

It was apparent that when the teachers talked about geography they frequently referred to aspects such as topography, climate or vegetation rather than other geographical aspects associated with population, culture or politics. In one lesson, a teacher used images to teach the concept of territory, showing various geographical spaces such as the Portada de Antofagasta, a natural arch located on Chile's coast (used to describe the concept of erosion), the Torres del Paine mountains, the Osorno volcano, the face of El Peñón mountain, and other images that alluded to natural spaces rather than human dynamics (Class observations ID38). Another teacher, also teaching the concept of territory, mentioned the territory's morphological characteristics and referred to Chile as "a long straight country" (Class observations ID18). A third teacher described Chile's geomorphology, mentioning what most teachers call the *four topographical layers* (Class observations ID17), i.e., Chile's four macro-relief units: the Andes Mountains, the Intermediate Depression, the Coastal Mountains and the Coastal Plains.

In their conceptualisation of geography, several of the teachers attributed a great deal of importance to physical geography and they understood territory as the space belonging to the state that contains the country's physical characteristics. This interchange emphasises how the teacher refers to the students' immediate surroundings when teaching about the natural elements found in geographical spaces, but did not refer to any of the local sociocultural constructions.

> Interviewer: "Are local surroundings used to teach geography?"
>
> Teacher: "To show what microclimates are perhaps"
>
> Interviewer: "But other types of things, the children's neighbourhood experiences, studying rural topics, are these things discussed in geography lessons?"

> Teacher: "Of course, analysing the characteristics of our surroundings, for example with Year 5 students I was teaching the characteristics of natural areas and from here we can see the Andes, which are an example of a mountain range, and the Coastal Mountains; we look at characteristics of the surrounding area in the classroom"
>
> (Interview ID12).

6.6.4 *Categories of Analysis, Geographical Concepts and Epistemological Paradigms of Geography*

In their lessons, many of the teachers emphasised the importance of natural resources as one of the fundamental elements of the geography curriculum. For example, one teacher planned an activity in which students built and painted a clay model showing Chile's macro-relief features, using different colours to show the main topographical features of each zone and macro-relief unit (further evidence of the predominance of physical geography). Students were then asked to identify economic activities in each of the zones and label their models to show which natural resources are extracted in each zone and their geographical locations (Class observations ID4).

In another lesson, the topic of natural resources was taught in a very descriptive and enumerative way. The various natural resources that are extracted from different parts of the country were identified, highlighting copper as Chile's primary natural resource. The existence of the 200 nautical mile exclusive economic zone was highlighted as important for the country's fishing industry (Class observations ID36).

Some teachers also acknowledged that although they wanted to teach the subject-space relationship, the curriculum is limited by a physical, unidirectional and at times deterministic focus. This was most apparent in two situations: the use of political maps and the elements of natural and anthropogenic landscapes.

By working with political maps students learn about the existence of different continents and countries. This is taught from a descriptive and enumerative perspective that disregards the social issues involved in these territorial configurations: "...the lesson starts with the teacher showing the students a political map of Latin America. The teacher asks a Year 1 student: Why is this political? The student replies that it is because it shows countries" (Class observations ID7). The opportunity to look more closely at the relationship between concepts such as territory, society, political organisation and countries was not exploited. Afterwards the teacher asked the students to identify the different Latin American countries by their colours on the map. Then:

> "What is the capital of Chile?" The teacher asks and then begins to read out or indicate the points on the map that represent the capital cities of each country. The teacher then asks a student "Demian, where is Peru and what its capital city is?" –points at the map. Demian replies "Lima"; "Claudia, what is the capital of Brazil?" "Brasilia" – Claudia replies. The teacher continues asking the same questions about the other countries (Class observations ID7).

With regard to the second situation, i.e., recognising elements of natural and anthropogenic landscapes and their relationships, overall the teachers did a satisfactory job of teaching the elements of landscapes and differentiating between natural and anthropogenic elements. However, in most cases they failed to explore the potential relationships that may exist within a single geographical space. The module goes on to describe elements of the landscape. The class do this by looking at the elements around them; some students look out the window and others look out the classroom door to identify elements, then they return to their seats and discuss which elements are natural and which are cultural, and then they draw them. This would have been an ideal opportunity to examine the relationships between subjects and space, however, the teacher did not explore the relationships between natural and cultural elements or between the subjects and geographical space. The teacher did not point out that the trees near the school are examples of exotic vegetation (insignis pine and eucalyptus) which are part of forestry plantations (as an example of an anthropogenic element). In other words, they would have been able to establish clear links between nature and society. They could have also made links to other relevant aspects such as the environmental problems affecting the locality, the scarcity of water resources or rural depopulation caused by these forestry plantations.

6.6.5 Categories of Analysis and Geographical Concepts, Methodological Models and Social Value of Geographical Knowledge

The teachers recognise the local context as a space in which geographical meanings exist and which is therefore relevant to teaching geography. One teacher, born in a rural area, describes it as a geographical space with few opportunities. She moved to the city to attend university but returned to her original locality to work as a teacher. She mentioned that when she is teaching she often remembers her own experience as a student and tries to plan her lessons in a way that gives her students the opportunities she didn't have:

> I didn't have this experience but I want children today to experience this, so that's what drew me to work here and where I worked previously; there are 14 localities in this municipality and I have worked in eleven of them (Interview ID44).

Later the same teacher added:

> …the thing that made me decide to stay here was that I felt that growing up in this rural area I was at a disadvantage compared to the city where I lived [referring to the city where she went to university]; there was a lack of knowledge, of everyday situations (Interview ID44).

By explaining the types of disadvantages of the locality where she grew up, the teacher expresses a sense of responsibility, valuing the teaching of geography as a means of redefining her spatial experiences. Following this same logic, the teachers considered it important to teach the local context first and then continue on to other levels. They believe that geographical meanings result from experiences within one's context. One teacher said, "There's no point in children being able to locate the Aztecs, the Diaguitas, if they have no idea where the Chilean Altiplano is, for example" (Interview ID21). Another teacher stated:

> We have a premise for working: we work from near to far. First we learn about what is nearby, then we gradually begin learning about things further away, because how can I talk about Miami to a child who hasn't ever seen the sea? (Interview ID23).

Other teachers also confirmed the relevance of the local context in constructing geographical meanings, but in terms of the way a society constructs its heritage. Two teachers recalled classes in which they asked students to draw something related to heritage and the first thing the children drew was the Chilean flag. The teachers' response was:

> We need to expand the vision. Heritage is more than a Chilean flag, it's more than the national anthem. It needs to be given a local meaning so that they feel that they are part of this identity. So it is something that needs to be worked on, gradually changing the social paradigm not only within the child; this issue is ingrained in society. People associate heritage with cuecas (Chilean folk dance), empanadas (typical Chilean pasty), the Chilean flag, but they don't associate heritage with anything personal, local, they don't feel that they are part of it (Interview ID45-46).

For these teachers, the local context is significant for students in terms of constructing geographical meanings that are linked to identity, roots and an understanding of geographical reality that originates from the spaces with which we are most familiar.

6.7 Conclusion

This chapter has identified that developing geographical thinking in Chilean schools is influenced by a complex range of inter-related factors including the quality of teacher geography knowledge, their different interpretations of geography knowledge and their application of these in classroom practices. At least three main sources of geographical knowledge for teachers were identified. Figure 6.4 outlines these. Despite their identification, it is clear to us that there is often very little connection made by teachers between these sources. Instead not only because they are highly disassociated or because there is no feedback among them, but because there is no stated purpose, tools (initial training, national curriculum) and significance (spatial experience) to structure geographical thinking.

In our discussions with teachers, five key characteristics of their practice in developing geographical thinking were identified.

First, teachers' understanding of geography and geographical knowledge is based almost exclusively around chorographic and topological localization.

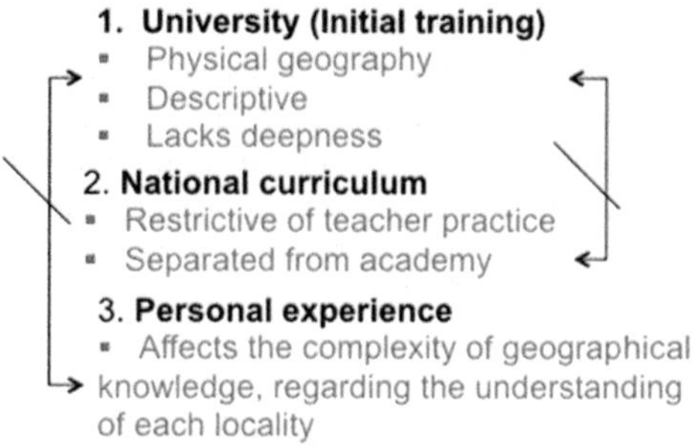

Fig. 6.4 Three different sources of geographical knowledge (*Source* Prepared by the authors)

Second, in connection with this, teachers tend to associate geography with the physical aspects of their surroundings, their country and the world with regards to climate, topography, flora and other features but not always in an integrated or systematic way. This is not to say that the teachers taking part in the research were not willing to teach about the relationships between human subjects and space, but rather that often these relationships were taught in a rather unidirectional and deterministic way.

Third, we would argue that teachers' knowledge of geography and geographical tools resembles the early stages of the evolution of geographical thinking. By tools we essentially mean conceptualisations and methodologies, although these are based on their epistemological origins. By early stages, we mean the early period of "scientific geography".

Fourth, the teachers associated geography with natural resources, the location of those resources and economic exploitation. But many of the teachers did not "bring" into the classroom the economic-spatial processes that occur in their immediate local contexts even though they had a good understanding of these processes themselves.

Fifth, it was apparent in the research that for many of these teachers, their personal experiences of space are redefined when they teach and this becomes a driving force towards a more complex geography education. However, this wealth of spatial experiences and the understanding that comes from personal experience is rarely linked to the school curriculum. It is possible that the way the national curriculum is designed and delivered imposes a limited and superficial vision of geography and that teachers' lack of knowledge of geography as a subject precludes them from actively discussing the curriculum; or possibly the teachers fail to recognise these experiences and understandings as geographical and are therefore incapable of generating a specific type of geographical thinking.

In conclusion, the kind of geographical thinking that can be developed by students is strongly impacted by the theoretical and conceptual abilities of their teachers when giving geography its own identity through a specific and recognizable knowledge—something that is not properly addressed in basic training. Moreover, this would be influenced by the underestimation of personal and common knowledge of space and its geographical character. Taking these results into account, it is possible to state that both, teachers and students, have a spatial knowledge, developing a geographical thinking of a local and experience-based nature that is not acknowledge as such; neither by themselves nor by the official curricula.

References

Alexandre, F. (2009). Epistemological awareness and geographical education in Portugal: The practice of newly qualified teachers. *International Research in Geographical and Environmental Education, 18*(4), 253–259.

Arenas, A., & Salinas, S. (2013). Giros en la Educación Geográfica: Renovación de los geográfico y lo educativo. *Revista Geográfica Norte Grande, 56*, 143–162.

Bahbahani, K., Case, R., Tu Huynh, N., & Sharpe, B. (2010). *Teaching about geographical thinking: A professional resource to help teach six interrelated concepts central to students' ability to think critically about geography*. Canada: Simon Fraser University Library.

Berger, P., & Luckmann, T. (1996). *La construcción social de la realidad*. Buenos Aires, Argentina: Amorrortu.

Biddulph, M., Lambert, D., & Balderstone, D. (2015). *Learning to teach geography in the secondary school: A companion to school experience*. London, UK: Routledge.

Bosque, J., & Ortega, F. (1995). *Comentario de textos geográficos: Historia y crítica del pensamiento geográfico*. Barcelona; España: Oikos-Tau.

Capel, H. (2012). *Filosofía y ciencia en la geografía contemporánea: Una introducción a la Geografía*. Nueva Edición Ampliada, Edición: Del Serbal; Barcelona; España.

Fenstermacher, G. D. (1994). The knower and the known: The nature of knowledge in research on teaching. *Review of Research in Education, 20*, 3–56.

Flick, U. (2004). *Introducción a la investigación cualitativa*. Madrid, España: Serie Educación Crítica, S. Morata y Fundación Paideia Galiza.

Gallestegui, J., & Galaz, J. (2009). Espacios para una geografía social, humanista y crítica. Primera edición. Valparaíso: Universidad de Playa Ancha de Ciencias de la Educación.

Garrido, M. (Ed.). (2009). *La espesura del lugar. Reflexiones sobre el espacio en el mundo educativo. Colección Investigación*. Santiago, Chile: Universidad Académica Humanismo Cristiano.

Giddens, A. (1997). *Las nuevas reglas del método sociológico: Crítica positiva de las sociologías comprensivas* (2nd ed.). Buenos Aires: Amorrortu.

Gómez, J., Muñoz, J., & Ortega, N. (1988). *El pensamiento geográfico: Estudio interpretativo y antología de textos de Humboldt a las tendencias radicales* (2a ed.). Madrid, España: Alianza.

González, I. (2002). El conocimiento geográfico e histórico educativo: La construcción de un saber científico. En la geografía y la historia, elementos del medio. Madrid: Ministerio de Educación y Ciencia, Online: www.ub.edu/histodidactica. 05 January 2011.

Gurevich, R. (2005). *Sociedades y territorios en tiempos contemporáneos: una introducción a la enseñanza de la geografía. Serie Colección Educación y Pedagogía* (1a. ed.). Fondo de Cultura Económica. Buenos Aires. Argentina.

Harvey, D. (1983). *Teorías, leyes y modelos en geografía*. Madrid: Alianza Editorial (trad. Gloria Luna Rodrigo).

Lacoste, Y. (1981). *"Geografía". Tomo IV. En F. Chatelet. (Dir): Historia de la Filosofía*. Madrid: Espasa Calpe.

Lidstone, J., & Williams, M. (Ed.) (2006). *Geographical education in a changing word*. (105) Dordrecht, The Netherlands: Springer.

Lincoln, Y., & Guba, E. (1985). *Naturalistic inquiry. Publicado por SAGE*. Newbury Park London New Delhi: International Educational and Professional Publisher.

McKernan, J. (1999). *Investigación-acción y curriculum. Métodos y recursos para profesionales reflexivos*. España: Morata.
Miles, M., & Huberman, A. M. (1994). *Qualitative data analysis: An expanded sourcebook* (2nd ed.). London: SAGE, International Educational and Professional Publisher.
MINEDUC. (1999). *Evaluación del Programa de Educación Básica Rural 1992–1998*. Santiago, Chile.
MINEDUC. (2012). Portal Red Maestros de Maestros. En línea. http://ww.rmm.cl/website/index.php?id_seccion=21, consultada: 15/01/2012 y 30/01/2013.
MINEDUC. (2013). Programa de Microcentro Rurales. http://www.convivenciaescolar.cl/index1_int.php?id_portal=50&id_seccion=4078&id_contenido=18587, consultada: 07/04/2013.
MINEDUC. (2014). Historia, Geografía y Ciencias Sociales, Cuaderno de Trabajo, Geografía. Módulo didáctico para la enseñanza y el aprendizaje en escuelas rurales multigrado. Geografía I, Clase 1. 1° a 6° Básico. Santiago, Chile.
Moraes, A. (2006). *Geografía, pequeña historia crítica*. GEOUNTREF-EDUNTREF: Buenos Aires.
Moreira, R. (2006). *Para onde vai o pensamento geográfico? Por uma epistemologia crítica*. Segunda Edición, São Paulo: Contexto.
Nogué, J. (2007). El Paisaje como Construcción Social. In J. Nogué (Ed.). *La construcción social del paisaje* (pp. 11–24). Madrid: Editorial Biblioteca Nueva España.
Nogué, J., & Romero, J. (2006). Otras geografías, otros tiempos. Nuevas y viejas preguntas, viejas y neuvas respuestas. In J. Nogué & J. Romero (Eds.), *Las otras Geografías* (1a. ed. pp. 15–50). Tirant lo Blanch: Valencia; España.
Palacio, J. L. (2011). Los estudios de Geografía en las universidades de América Latina; desarrollo, situación actual y perspectivas. *Investigaciones Geográficas, 74*, 107–124.
Prats, J. (2003). *Líneas de investigación en didáctica de las ciencias sociales En: História & Ensino Revista do Laboratório de Ensino de Históriória/UEL* (Vol. 9). Brasil: Universida de Estadual de Londrina.
Reinfried, S., Schleicher, Y., & Rempfler, A. (Eds.) (2007). Geographical views on education for sustainable development. *Proceedings of the Lucerne-Symposium, Switzerland, July 29–31, Geographiedidaktische Forschungen* (Vol. 42, pp. 243–250).
Republic of Botswana. (2000). Botswana General Certificate of Secondary Education Teaching Syllabus—Geography. Curriculum Developing Division Department of Curriculum Development and Evaluation, Ministry of Education.
Santos, M. (1990). *Por una geografía nueva. Serie Espasa. Universidad, 20*. Espasa-Calpe: Madrid. España.
Schwandt, T. (1998). Constructivist, interpretivist approaches to human inquiry. In N. Denzin & Y. Lincoln (Eds.), *Handbook of qualitative research* (pp. 118–137). London: Sage.
Shulman, L. S. (1993). Renewing the pedagogy of teacher education: The impact of subject-specific conceptions of teaching. In L. Montero & J. Vez (Eds.), *Las didácticas específicas en la formación del profesorado* (pp. 53–69). Madrid: Síntesis.
Stake, R. M. (1998). *Investigación con estudios de casos*. Madrid, Morata: España.
Stake, R. M. (2000). Case studies. In N. Denzin & Y. Lincoln (Eds.), *Handbook of qualitative research* (2nd ed., pp. 435–454). California: Sage Publications.
The Critical Thinking Consortium. (2010–2011). *Resources for critical thinking*. Canada: Vancouver.
Tuan, Y.-F. (2007). *Topofilia: un estudio de las percepciones, actitudes y valores sobre el entorno*. España: Melusina.

Chapter 7
Geographical Thinking and Its Role in Climate Change Education: The Case of Singapore

Chew-Hung Chang and Liberty Pascua

7.1 Introduction

Geographical thinking affords excellent avenues for teachers to help students engage with the issue of climate change. As an important theme in the investigation of human-environment interactions, climate change in geographic education emphasises the building of core understanding of climate science (Dalelo 2011). In order to comprehend the causes, impact and human management of the phenomenon, students also require understanding of the concepts of space, place, scale, and movement, in addition to human-environment interaction.

The following excerpt illustrates how crucial geographical thinking is to learning about climate change:

> There is an argument about climate change that goes like this. The UK's contribution to global emissions of greenhouse gas is only a small percentage. There's not much point in taking responsibility for our own place when India and China are growing as they are. Now I might have found that a comforting argument but it seems it's a totally inadequate geography. What that small percentage counts is the greenhouse gas emissions from the United Kingdom directly. In that sense it treats the UK as an isolated entity - but it is not. That calculation it seems, misses out the effect of all the things we import from elsewhere, many of them indeed from China. We demand of those goods, that we do not count as our own, the pollution of producing them. (Massey 2007, 1)

Without the concepts of space, scale, movement and human-environment interaction, it would be impossible to appreciate that climate change is a multi-faceted and complex issue. If students were only presented with a graph of the relative emissions of greenhouse gases by countries, it would need deeper geographical thinking in order for them to critically engage the source and ask the same questions that Professor Doreen Massey did in her radio interview. Students may

C.-H. Chang (✉) · L. Pascua
National Institute of Education, Nanyang Technological University, Singapore
e-mail: chewhung.chang@nie.edu.sg

C. Brooks et al. (eds.), *The Power of Geographical Thinking*,
International Perspectives on Geographical Education,
DOI 10.1007/978-3-319-49986-4_7

not be renowned Geographers like Professor Massey, but a teacher who has deep understanding of geographical concepts has the opportunity to provide the graph to students, and to guide them by asking questions such as "Where was the shirt you are wearing made?" and "Do you think that the carbon emissions producing that shirt are included in this graph?" Hence, the teacher plays a vital role in guiding the students' geographical thinking.

Therefore, the central enquiry question that overarches the discussion in this chapter is "How can geographical thinking help students understand climate change?"

Humans have a key role to play in shaping our lived environments in the Anthropocene (Crutzen and Stoermer 2000). Environmental and accompanying societal changes are happening so rapidly that children today are inundated with information across mainstream as well as social media (Ambusaidi et al. 2012), making it necessary for them to engage critically with the vast volume and diversity of knowledge that they encounter. There is also room for critical engagement in terms of questioning the perceived contradictory discourses of climate change. Much has been said about the importance of climate change literacy as a critical life-skill (Anderson 2012; Harrington 2008; Kagawa and Selby 2012; McCaffrey and Buhr 2008). However, the nature of climate change literacy is complex. In addition, "uncertainty" and "complexity" pepper the discourse on climate change, making its teaching and learning complex.

With regards to helping teachers frame there teaching about climate change, the Future 3 curriculum (Young and Muller 2010) is very useful to consider. As the term implies, Future 3 is one of the three curriculum models posited based on a review of the relationship between knowledge taught in school and the learner's everyday experiences. The school and non-school boundaries of knowledge are described in 3 scenarios namely:

- Future 1—boundaries as given
- Future 2—a boundary-less world
- Future 3—boundary maintenance as a condition for boundary crossing.

In the Future 1 curriculum, transmission of knowledge is the main modality of imbuing children with geographical thinking to engage issues in society. Being able to read and understand news about climate change is a necessary condition to succeed in a Future 1 curriculum, but it is doubtful that such is sufficient for our current world. There are often confounding and contradictory messages published on print media. The learner needs to be able to question the validity of the source and evaluate what he or she is reading.

In the Future 2 curriculum, there is an over-socialisation of knowledge. While the school provides the key factual knowledge about the issue, the learner is left unsatisfied with the knowledge given as he or she is unable to make sense of the volume of information available through the media using this factual knowledge. The learner may become overly sceptical, criticising and refuting all knowledge that is taught in the classroom, very likely referring to their own constructed naive theories. The danger is, of course, a sceptical treatment of knowledge that is provided at school, leading to a potential dismissal of the relevance of school knowledge.

The Future 3 curriculum envisages a future in which a learner engages with the information encountered, within the contextual understanding of school knowledge, and asks critical questions that lead to the development of deeper understanding of the issue at hand. Future 3 is referred to here as the "Goldilocks solution", such that it veers away from the extremes of knowledge with fixed boundaries on one end, and knowledge with no disciplinary demarcations on the other. In the Future 3 curriculum, geographical thinking when well taught, provides a lens for inquiry so that the student is able to see the relevance of acquired knowledge to society. Clearly, the key stakeholder that is largely responsible for facilitating the knowledge transaction between geographical thinking and the students' engagement with information is the teacher. Hence it is important for teachers to understand students' geographical thinking, or the lack of, in teaching the topic of climate change within the Future 3 curriculum.

Nonetheless, the road to an F3 curriculum paving the way for a climate literate citizenry is a long and complicated one. The following sections tackle how knowledge, perceptions and attitudes of students in Singapore on the phenomenon become challenges to meaningful learning. The discussion that follows is based on observations from four years of research on climate change education in the city-state. Finally, it illustrates ways in which geographical thinking could or should be utilized to address the issues raised.

7.2 Misconceptions and Faulty Knowledge

Students' prior knowledge of a topic in geography can bar them from learning new concepts correctly (Dove 1999). Such naïve knowledge can be erroneous as the conception can be incorrect, incoherent or incomplete. With these misconceptions organized in mental models, they become well-entrenched and have the tendency to proliferate and persist (Liarakou et al. 2011).

The large quantities of information about climate change in the media has made children pessimistic and often conjure an image of a depressing future on a superheated earth (Jonsson et al. 2012; Tan 2013), with apocalyptic consequences (Özdem et al. 2014). This heightened awareness is not always accompanied by correct and deep understanding of the issue. Many misconceptions about climate change have been documented, including how Ozone depletion causes global warming (Osterlind 2005) and how ultraviolet heat is trapped that results in global warming (Lambert et al. 2012).

In a study conducted to find out if there are multiple dimensions to the misconceptions of students, students were interviewed to diagnose their baseline understanding of climate change. The semi-structured interview questions were patterned after the syllabus structure of Secondary 4 geography. The key questions asked were as follows:

1. How does climate change happen?
2. What are the natural causes of climate change?
3. What are human activities that exacerbate the climate change process?
4. What is the enhanced greenhouse effect?

Although a conscious decision was made to not ask direct questions that would lead students to voice alternative conceptions such as those already identified in the literature, most of the conversations landed on how 'the hole in the sky', for example, leads to global warming. Interestingly the research study produced six (Chang and Pascua 2015) representations of mental models that illustrate students' misconceptions about the greenhouse effect and its enhanced version. Although the scope of the examples is limited to understanding the greenhouse effect, exemplars are provided on how teachers can help students develop fuller understanding. Each of these representations will be discussed briefly to uncover the key misconceptions on understanding the greenhouse effect.

There is no distinction between pollutants and greenhouse gases from this mental model in Fig. 7.1. In fact, heat is only thought of as heat from the sun. Terrestrial radiation is missing from the model. Although the students have listed artificial sources as another origin of heat, confounding ideas such as pollutants that trap heat further complicate the model.

In Fig. 7.2, the same model of how heat is trapped by gases is perpetuated. The only difference is that the gases are now conceptualised as a distinct layer blanketing the earth. This unfortunate coincidence with the popular use of verb "blanketed" in a laypersons' description of greenhouse gases has not made it easier for students to understand the phenomenon. Terrestrial radiation is also missing from this mental model.

One common problem identified in the literature that was also apparent in the study was the notion of how the gap in the Ozone layer is the cause of global warming. While some students believed that the Ozone gas traps heat, most of those interviewed expressed that the Ozone reflects heat away from the Earth. The intensification of the heating process was attributed to the increase in the sun's

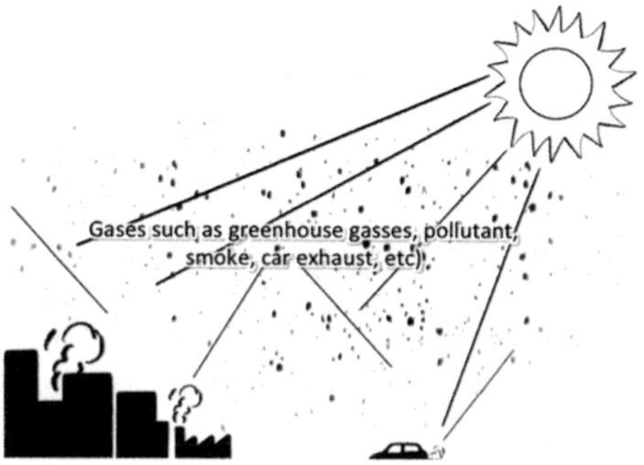

Fig. 7.1 The misconception that gases trap heat from the sun. *Source* Chang and Pascua (2015), p. 6

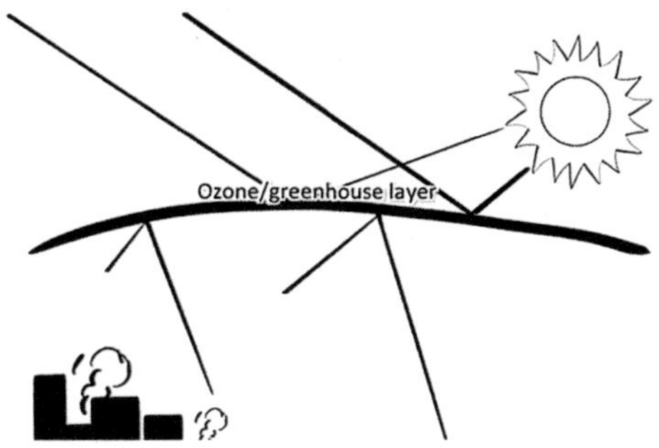

Fig. 7.2 A layer traps heat. *Source* Chang and Pascua (2015), p. 7

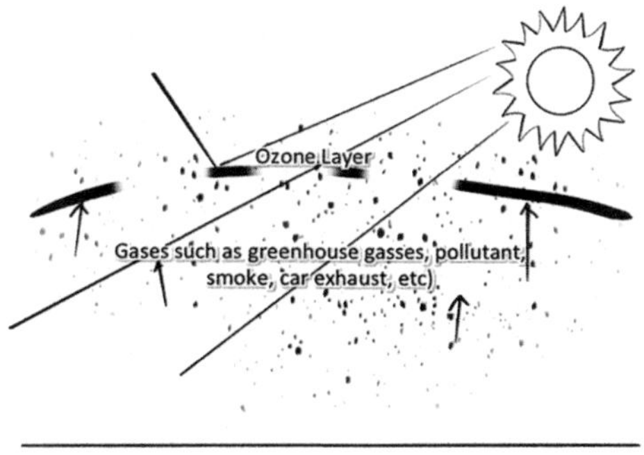

Fig. 7.3 Gases destroy the Ozone layer. *Source* Chang and Pascua (2015), p. 7

capacity to bombard the Earth with more damaging rays as the Ozone's blanket of protection is destroyed by CFCs, other greenhouse gases, pollutants and heat from man-made sources. Note that terrestrial radiation was also missing in the two mental models in Figs. 7.3 and 7.4.

The study showed that the misconceptions were largely built with coherent but incorrect and incomplete understanding. Students are unlikely to be aware of their misconceptions about the energy–atmosphere interaction, as there is consistency and coherence of their mental models in expanding the logic of how the greenhouse effect traps more heat. There is prevalence of confounding ideas, with elements injected into their arguments to substitute correct concepts, similar to what Shepardson et al. (2009, 2012) and Andersson and Wallin (2000) have found. Indeed, Andersson and Wallin (2000) argue that the Ozone-related models serve as a convenient and easy route to explain the warming process as 'the barrier thins out, more radiation can come through' (p. 1108).

Geographical thinking might help the students unravel the relationship between greenhouse gases and the longer time change in global average temperature. Catling's (2015) notion of inadequate geographies come to mind as students are unaware of the geographical process operating in larger time and spatial scales. While the literature has shown students' perception of industrial activities as

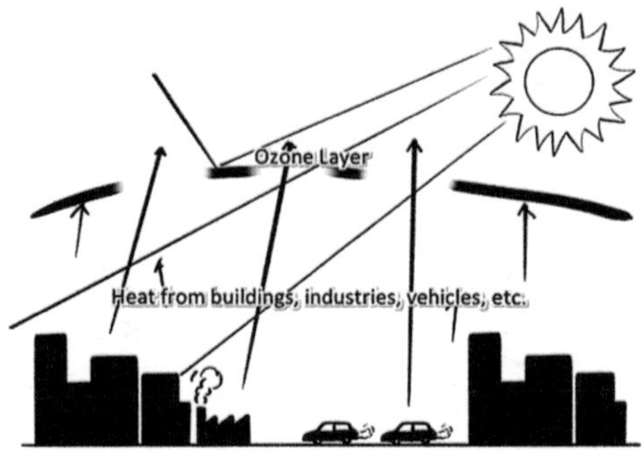

Fig. 7.4 Heat destroy the Ozone layer. *Source* Chang and Pascua (2015), p. 8

culprits of climate change, the understanding of the relationship between humans and the environment is shallow at best. Missing from the understanding is how energy from the sun interacts with the ground. There is also a simple treatment of the atmosphere as either composed of a blanket or layer of greenhouse gasses or an erroneous belief that the gasses themselves create a hole in the sky, allowing more heat to reach the earth. There is a gap in the understanding of movement of energy from the top of the atmosphere to the surface and then how this absorbed energy heats up the atmosphere. There is also the missing concept of time. Solar radiation needs time to reach the surface of the earth before it gets heated up. The stored heat then takes time to heat up the atmosphere. Clearly, the students are not aware of the passage of time but see the entire process as a snapshot in time, resulting in the absence of the heat from the ground in the mental models derived. While the interviews did not extend the questions further to explore the students understanding of how the interaction between the incoming solar radiation and the outgoing terrestrial radiation will differ from place to place, say between a low lying flat area near the tropics to a south facing slope on a mountain in a temperate zone in the Northern Hemisphere during summer, it is apparent that the students' understanding has not extended to these geographical concepts.

Although there has been efforts to incorporate the science of weather and climate in the formal curriculum (Dalelo 2011), geographical thinking will be instrumental in liberating students beyond their current misconceptions. However, there are other issues apart from misconceptions and naïve knowledge. Despite the heightened awareness about climate change among learners (Ambusaidi et al. 2012), there is a general pessimistic outlook of a future that they do not quite understand (Jonsson et al. 2012; Özdem et al. 2014; Tan 2013). While naïve knowledge inhibits individuals from engaging the climate change topic critically and intelligently (Bord et al. 2000), the general apathy towards the topic remain an issue for geography educators. The conundrum of getting students interested enough to learn about the topic before teachers can help them uncover and correct their naïve knowledge needs to be resolved.

7.3 Apathy, Assessment and Geographical Thinking

Among learners, the literature is indicative of confusion and doubt about one's role in environmental stewardship (Ahmad et al. 2012; Schreiner et al. 2005). Climate change is viewed with a cynical eye—difficult, overwhelming and beyond a young person's capacity for cognition and control (Kuo 2010). Conversations with students in Singapore, for instance, yielded indications of an apathetic deportment on climate change. While categorically declaring that climate change is an important and urgent issue, they were hesitant to agree on whether they should be actively taking part in efforts for mitigation and adaptation. The two most common reasons given by the students were:

1. We live in the 'safe zone'.
   (a) Singapore is shielded physically by Malaysia and Indonesia.
   (b) There are no natural disasters in Singapore.
   (c) It is the government's responsibility to do something about climate change. Singapore is already a garden city.
2. If it is not in the test, then I am not interested.

The responses point to a lack of appreciation for the geographical concepts of place and scale. For the first point, indeed students mainly vicariously experience natural disasters as they happen in other parts of the world. Being located just a degree north of the equator, and being sheltered to the west by Sumatra and to the east by the island of Borneo, Singapore's weather is in the doldrums—literally. The weather forecast varies minimally: temperatures throughout the year are high, the winds are weak save for the arrival of the northeast monsoon, and with short and intense local showers typically dousing the day's heat. The island is tectonically stable and has never been visited by a cyclone, except for a freak storm in 2001. It is ironic that although the Singapore government has demonstrated concerted efforts through the formation of the inter-ministerial National Climate Change Secretariat in tackling the issue, the students were all born in a society wherein state-dominance in social policies has inadvertently shaped the collective direction and aspirations of the nation. It was not surprising then that they referred to the rather long-running and successful government campaign of turning the city-state into a garden city as their reference in arguing that it is the government's responsibility to solve the climate change issue. Overall, these factors have helped students develop a strong sense of place about Singapore—a garden-state that is protected from natural disasters. What is unsettling about the responses is, of course, how unaware students are of their place in the entire global climate system. They are unaware of how each member of the global community has a part to play in climate change mitigation. Doreen Massey's excerpt shared in the preceding sections come to mind in which students are unaware of the imported and exported emissions that Singapore residents are responsible for. Indeed, this unawareness is most manifested in thinking that mitigating climate change should be the government's responsibility. The situation is not irreversible and teachers can use geographical thinking as a way to help students develop a more holistic understanding of their role in the larger global environmental system. However, an additional impediment presents itself even when teachers try to use geographical thinking to help students see that this is indeed their problem.

Assessment, while an indispensable component of the curriculum, could also obstruct learning. On one hand, proper assessment tools empower educators to determine the efficacy of pedagogical approaches and most importantly, to make certain that students are learning effectively. Assessment provides opportunities for the teacher to fine tune instruction (Drake 2007; Raggatt et al. 1985; Scott 2001). On the other hand, assessments can also pose challenges. From the interviews, it appeared that the students were most interested with climate change only if the topic is part of the written examinations. When asked why they are not interested in learning about climate change, students were quick to counter, "Is it included in the exams?" The students claim that they study only what they are supposed to learn for the exams. In fact, they complained that they have a lot of topics to cover and that climate change is just one of them. This observation is not uncommon in the education literature in Singapore (Cheah 1998; Higgitt 2006).

In combination, these issues present impediments to translating geographical knowledge into understandings about climate change. The chapter argues that the teachers as curriculum makers and enactors are central to the development of geographic thinking, to help students become more aware of their role in the human-environment relationship. The modality of assessment can inhibit the development of geographical thinking. If questions are phrased to solicit factual answers or explanation of processes, there is little scope for students to employ geographical thinking. Using the example of assessment in Geography in Singapore, there has been a clear move away from description and explanation question types to incorporate more critical and evaluative questions to encourage students to employ geographical thinking. A "levels marking" approach was introduced for part of the national written examination paper for geography in 2007 (Singapore Examinations and Assessment Board 2010) so as to assess students' ability to discuss and evaluate geographical problems. The introduction of levels marking was intended to prepare students "to meet the challenges of an increasingly globalised world … to promote critical and creative thinking skills, and to nurture problem-solving and independent learning abilities in students" (Sellan et al. 2006). As geographical thinking prepares the child to engage in the global issues of our time, this format of question requires the student to evaluate the issue and then construct an opinion, after analysing the information provided, not unlike the ideals of a Future 3 curriculum. Despite this change, the study still showed that the students only studied what they thought was included in the examinations. What students fail to understand is that topics are often interrelated. How does one understand about shoreline erosion independent of the context of sea-level rise or how does one understand the migration of environmental refugees in light of climate change? The topical treatment of school geography in this case will be detrimental to a holistic understanding of the human-environmental interaction.

Nevertheless, this points to an even more crucial need for the teacher as a curriculum maker to help students see the power of geographical thinking in understanding the topic of climate change.

7.4 Conclusions

In developing a case for teachers to use geographical thinking to guide students' learning about climate change, the first issue this chapter has tackled is to determine who our students are. In that, a description is given on the dire need for learners to engage with the information they encounter about climate change—neither to accept it uncritically nor to refute all that they come into contact with. They need to approach it from a middle ground where they raise questions and find answers by investigating them through geographical thinking. The teacher's role as a curriculum maker cannot be overemphasized as he or she needs to design learning activities, curate learning artefacts and pose meaningful and provocative questions to guide the students' geographical thinking.

However, teachers will first need to know what students typically misunderstand before they can guide and facilitate their learning. The key issues uncovered in this chapter include students' unrealised misconceptions, apathy towards the topic and preoccupation with standardised testing. A flawed mental model is normally built with coherent but incorrect and incomplete elements. Consequently, students use inadequate geographies and weave them into their personal theories or naïve conceptions about climate change. It is unlikely that students are able to detect the misconceptions, as they are often consistent and coherent but incomplete. The apathy that stems from an erroneous belief that they are safe and, again an incomplete understanding of their role in the global environmental system is equally worrying. The tension created by the need to cover topics for examinations has also been cited as a reason for disregarding the topic altogether. The common thread that ties these three key issues together is the lack of geographical thinking. The incomplete understanding of the causes of climate change stems from a lack of appreciation of the geographical concepts of space, time and scale. The apathy stems from a deeply entrenched context of place that is erroneous. While the physical environment of Singapore is perceived as being sheltered from natural disasters, the reality is that when the global climate system changes, daily weather events will change dramatically across different places on earth. More importantly, the Singapore student is not absolved from the responsibility to mitigate climate change even if he or she is in a "safe zone". Much of the greenhouse emission of production and consumption of resources for this Singapore student has been imported or exported but these are not included in the official figures. Finally, there is also the unrealised understanding that all topics in the examination are interrelated. Teachers are presented with opportunities to demonstrate how geographical thinking will help connect the topics and consequently help them do better in the examinations.

Geographical thinking is the key to help students uncover their misconceptions and to help them understand the topic of climate change geographically and holistically. However, this approach assumes that the teacher is able to engage geographical thinking in interpreting the curriculum documents, curate relevant resources for instruction and design meaningful assessment tasks to help students learn. The truth is that different teachers are at different stages of readiness to use geographical thinking, depending on their own experiences when they learned geography. Professional development in the form of conferences and workshops can help teachers stay updated on the latest development in geography. Consequently this can help teachers use geographical thinking in their classrooms. However, there is also a need to identify best practices and case studies so that teachers can reflect on how other teachers are using geographical thinking, and to develop their own craft to harness the power of geographical thinking, especially in teaching climate change.

Acknowledgements This paper refers to data from the research project "The Hole in the Sky is causing global warming?: Restructuring Prior Knowledge in Climate Change Education through Conceptual Change" (RS 5/12 CCH) funded by the Research Support For Senior Academic Administrator (RS-SAA) Grant, National Institute Of Education (Shepardson, et al.). The views expressed in this paper are the authors' and do not necessarily represent the views of NIE

References

Ahmad, A. L., Rahim, S. A., Pawanteh, L., & Ahmad, F. (2012). The understanding of environmental citizenship among malaysian youths: A study on perception and participation. *Asian Social Science, 8*(5), 85. doi:10.5539/ass.v8n5p85

Ambusaidi, A., Boyes, E., Stanisstreet, M., & Taylor, N. (2012). Omani students' views about global warming: Beliefs about actions and willingness to act. *International Research in Geographical and Environmental Education, 21*(1), 21–39. doi:10.1080/10382046.2012.639154

Anderson, A. (2012). Climate change education for mitigation and adaptation. *Journal of Education for Sustainable Development, 6*(2), 191–206.

Andersson, B., & Wallin, A. (2000). Students' understanding of the greenhouse effect, the societal consequences of reducing CO_2 emissions and the problem of ozone layer depletion. *Journal of Research in Science Teaching, 37*(10), 1096–1111. doi:10.1002/1098-2736(200012)37:10<1096:aid-tea4>3.0.co;2-8

Bord, R. J., O'Connor, R. E., & Fisher, A. (2000). In what sense does the public need to understand global climate change? *Public Understanding of Science, 9*(3), 205–218.

Catling, S. (2015). What geographical knowledge might early years/primary teachers have? In Paper presented at the Making an impact—Geographical Association Annual Conference 2015, Manchester.

Chang, C.-H., & Pascua, L. (2015). Singapore students' misconceptions of climate change. *International Research in Geographical and Environmental Education*, 1–13. doi:10.1080/10382046.2015.1106206

Cheah, Y. M. (1998). The examination culture and its impact on literacy innovations: The case of Singapore. *Language and education, 12*(3), 192–209.

Crutzen, P. J., & Stoermer, E. F. (2000). The 'anthropocene'. *Global Change Newsletter, 41*, 17–18.

Dalelo, A. (2011). Global climate change in geography curricula for Ethiopian secondary and preparatory schools. *International Research in Geographical and Environmental Education, 20*(3), 227–246. doi:10.1080/10382046.2011.588505

Dove, J. (1999). *Immaculate misconceptions*. Sheffield: Geographical Association.

Drake, S. M. (2007). *Creating standards-based integrated curriculum: Aligning curriculum, content, assessment, and instruction*. Thousand Oaks, California: Corwin Press.

Harrington, J. (2008). Misconceptions: Barriers to improved climate literacy. *Physical Geography, 29*(6), 575–584.

Higgitt, D. (2006). Finding space for education for sustainable development in the enterprise economy. *Journal of Geography in Higher Education, 30*(2), 251–262.

Jonsson, G., Sarri, C., & Alerby, E. (2012). "Too hot for the reindeer"—voicing Sámi children's visions of the future. *International Research in Geographical and Environmental Education, 21*(2), 95–107. doi:10.1080/10382046.2012.672668

Kagawa, F., & Selby, D. (2012). Ready for the storm: Education for disaster risk reduction and climate change adaptation and mitigation. *Journal of Education for Sustainable Development, 6*(2), 207–217.

Kuo, S.-Y. (2010). *The public understanding of climate change: A case study of taiwanese youth*. New Jersey Institute of Technology, Department of Chemistry and Environmental Science.

Lambert, J. L., Lindgren, J., & Bleicher, R. (2012). Assessing elementary science methods students' understanding about global climate change. *International Journal of Science Education, 34*(8), 1167–1187. doi:10.1080/09500693.2011.633938

Liarakou, G., Athanasiadis, I., & Gavrilakis, C. (2011). What Greek secondary school students believe about climate change? *International Journal of Environmental and Science Education, 6*(1), 79–98.

Massey, D. (2007). *Doreen Massey on the importance of geography—transcript radio 4 today programme*. Gaian Economics. Retrieved from http://gaianeconomics.blogspot.co.uk/2006/12/doreen-massey-on-importance-of.html

McCaffrey, M. S., & Buhr, S. M. (2008). Clarifying climate confusion: Addressing systemic holes, cognitive gaps, and misconceptions through climate literacy. *Physical Geography, 29*(6), 512–528.

Osterlind, K. (2005). Concept formation in environmental education: 14-year olds' work on the intensified greenhouse effect and the depletion of the ozone layer. *International Journal Of Science Education, 27*(8), 891–908. doi:10.1080/0950069050038264

Özdem, Y., Dal, B., Öztürk, N., Sönmez, D., & Alper, U. (2014). What is that thing called climate change? An investigation into the understanding of climate change by seventh-grade students. *International Research in Geographical and Environmental Education, 23*(4), 294–313. doi:10.1080/10382046.2014.946323

Raggatt, P. C. M., Weiner, G., & Open, U. (1985). *Curriculum and assessment: Some policy issues: A reader*. New York: Pergamon Press, in association with the Open University.

Schreiner, C., Henriksen, E. K., & Kirkeby Hansen, P. J. (2005). Climate education: Empowering today's youth to meet tomorrow's challenges. *Studies in Science Education, 41*(1), 3–49. doi:10.1080/03057260508560213

Scott, D. (2001). *Curriculum and assessment* (Vol. 1). Westport, Connecticut: Ablex Publishing.

Sellan, R., Chong, K., & Tay, C. (Producer). (2006). *Assessment shifts in the Singapore education system*. Singapore Examinations and Assessment Branch. Retrieved from http://www.iaea2006.seab.gov.sg

Shepardson, D. P., Choi, S., Niyogi, D., & Charusombat, U. (2012). Students' conceptions about the greenhouse effect, global warming, and climate change. *Climatic Change, 113*(3–4), 1097–1097. doi:10.1007/s10584-012-0472-y

Shepardson, D. P., Niyogi, D., Choi, S., & Charusombat, U. (2009). Seventh grade students' conceptions of global warming and climate change. *Environmental Education Research, 15*(5), 549–570. doi:10.1080/13504620903114592

Singapore Examinations and Assessment Board. (2010). GCE O-level syllabuses—combined humanities 2192 Singapore Examinations and Assessment Board.

Tan, G. C. I. (2013). Voices in Singapore: Young people visioning their futures. In M. Robertson & S. Tani (Eds.), *Young people: Cross-cultural views and futures* (pp. 107–124). Victoria, Australia: Acer Press.

Young, M., & Muller, J. (2010). Three educational scenarios for the future: Lessons from the sociology of knowledge. *European Journal of Education, 45*(1), 11–27.

Chapter 8
Geographic Education for Sustainability: Developing a Bi-national Geographical Thinking Curriculum

Fabián Araya Palacios, Alex Oberle, Ximena Cortés Quezada and Mollie Ullestad

8.1 Introduction

The most pressing global issues are complex and multi-faceted, defying simple solutions and requiring action on multiple fronts. Such problems include, for example, climate change, terrorism and geopolitical instability, and famine and food security. A significant factor in each of these is the availability of natural resources: the intensive use of fossil fuels hastening global warming, the scramble for resources contributing to a civil war or conflict, and a lack or uneven distribution of food causing a famine. With a global population of over 7.2 billion and increasing resource consumption in middle income countries, the concept of sustainability has risen to the forefront:

> Everything that we need for our survival and well-being depends, either directly or indirectly, on our natural environment. Sustainability creates and maintains the conditions under which humans and nature can exist in productive harmony, that permit fulfilling the social, economic, and other requirements of present and future generations (EPA 2014).

The bi-national project described here develops geographic thinking by employing sustainability as a means of incorporating physical and human geography. Enlisting the support of Chilean and North American pre-service geography/social science teachers, we developed a series of standards-based lesson plans and associated curricular materials on the theme of sustainability that link the Coquimbo region of Chile with the state of Iowa in the United States. While the

F.A. Palacios (✉) · X.C. Quezada
University of La Serena, La Serena, Chile
e-mail: faraya@userena.cl

A. Oberle
University of Northern Iowa, Cedar Falls, USA

M. Ullestad
University of Missouri, Columbia, USA

C. Brooks et al. (eds.), *The Power of Geographical Thinking*,
International Perspectives on Geographical Education,
DOI 10.1007/978-3-319-49986-4_8

geography of the two areas differs substantially, there are commonalities which demonstrate the global nature of sustainability and the fact that in any given region or country there are issues that are occurring simultaneously across the globe. For this project, common themes include aspects of sustainability, urbanization, agriculture, energy use, pollution, and other examples. As such, this sustainability focus demonstrates the power of geographic thinking by articulating with the work of Jackson (2009) and the GeoCapabilities initiative.

This chapter employs sustainability as a focus which is used to explore how physical and human geography develop geographical thinking. We begin with an overview of geographic thinking, including different ways in which it is defined. Next, since this is a bi-national project, we describe geographic thinking in the Chilean context, comparing it to the North American perspective. We then devote a considerable portion of our chapter to describing the project in detail, including showcasing examples of the lesson plans that are central to the endeavour. Following this, we detail the importance of sustainability in linking physical and human geography, including how the concept connects the two halves of the discipline to develop geographic thinking. Since an education course and pre-service teachers were a central part of this project, there is also a discussion regarding how teachers can support geographic thinking. Finally, the chapter ends with a conclusion that includes directions for future research and collaboration.

8.2 Geographical Thinking

Geographical thinking encompasses perspectives, skills, and concepts that are needed to analyse and explain our world. Yet before elaborating on geographic thinking, it is important to first understand geo-literacy—the basic knowledge that one needs to know about the world. Geo-literacy consists of three things: interactions, interconnections, and implications (Edelson 2011). Interactions are how the world works in terms of systems, both human and physical. Interconnections describe how the world is connected through these systems, with the idea that something happening in one locale not only has an effect on a neighbouring location, but can also impact faraway places. Implications involve the decision making process, understanding how local, regional, and global decisions have an impact, often one that is far-reaching in terms of both space and time.

Relating to geo-literacy, but extending into geographic thinking, is the idea that geography involves particular perspectives that either do not exist in other disciplines or are tangential to those fields. The Rediscovering Geography Committee of the National Research Council suggests three perspectives that represent a geographer's ways of looking at the world (National Research Council 1997). These perspectives form part of a matrix that underscores geography as a discipline that excels in synthesizing data. The perspectives are *integration in place*, the idea that geographic places are a sort of laboratory for understanding processes and phenomena; *interdependencies between places*, which focuses on flows of people, ideas, and products that link places to one another; and lastly, and perhaps most significantly, a unique geographic perspective is *interdependencies among scales*, where geographers can zoom in or out, analysing how local, regional, national, and global scales are linked.

Building on this foundation of geo-literacy and geographic perspectives is geographic thinking, arguably best articulated by Jackson (2006). Jackson identifies four concepts that are inherent to thinking geographically and that undergird our project. First, is *space and place* where space and the spatial perspective is abstract, while place is a humanized version of space. Next, is *scale and connection* where scale is referenced in terms of connections among scales; local decisions can have global consequences and global policies and actions have a differential effect on various local areas. The concept of *proximity and distance* follows, but rather than in a traditional sense of measured distance, Jackson defines this as a socially-constructed and dependent on technologies and other means that shrink actual distance. Finally, *relational thinking* focuses on differences and similarities and is firmly in the realm of human geography, and is therefore not as directly applicable to certain aspects of sustainability.

The idea of GeoCapabilities is a conceptualization that demonstrates the power of geographical thinking. Building on the notion of *human capabilities*, this concept enlists geography to prompt people to take responsibility for their lives by gaining the specialized knowledge to become active, engaged citizens. The topic of sustainability is an ideal match with GeoCapabilities because it integrates the natural and human world, is evident and actionable at all scales from local to global, and encourages people to make informed decisions that have both short term and long term consequences. As such, the bi-national curricular materials that comprise this project demonstrate this concept as it encompasses the three essential components: geographical thinking, choices teachers make, and student experiences (GeoCapabilities 2015). While the GeoCapabilities approach and process was not employed in developing the bi-national project, many of its core ideas are present and could serve to further leverage the impact of this project.

8.3 Geographical Thinking in Chile

> Thinking geographically is a uniquely powerful way of seeing the world. While it does not provide a blueprint, thinking geographically does provide a language, a set of concepts and ideas, that can help us see the connections between places and scales that others frequently miss. That is the power of thinking geographically (Jackson 2009, p. 9).

In this project, thinking geographically centres on interdependence in the way that the Geographical Association describes as 'linking the physical world and human environments and understanding the concept of sustainable development" (Geographical Association 2009, p. 11).

In Chile, the interest in defining geographical thinking began when the *Sociedad Chilena de Ciencias Geográficas* (Chilean Society of Geographic Science) made a national public statement in June 2013 about the lack of explicitness of the geographic content in the school curriculum. This raised much concern, and the Chilean Society later repeated this concern to the Social Sciences Team of the Curriculum and Evaluation Unit of the Ministry of Education (MINEDUC).

Currently, after asking for further details from the key individuals involved, the Ministry of Education in Chile is undertaking curricular changes in several areas and subcategories of teaching, with the purpose of modifying the current curriculum. This measure will bring about positive changes in the field of geographic education and, above all, will introduce geographic thinking into the school curriculum (Butt 2011). During 2014 and 2015, the Social Sciences Team has worked to incorporate geographic contents and skills into the different levels of the Chilean national curriculum.[1] The main observations and conclusions are as follows: the need for more of a presence of geographic thinking in school teaching and a clear sequence to its learning; addressing the current weak approach to geographic thinking in the curriculum and a lack of significant geographic themes in academic training and preparation; and the absence of high quality geographic skills in the geographic thinking of students.

In order to overcome the challenges noted in these observations and conclusions that were expressed by an array of members of the educational system, the curricular changes are currently (at the time of writing in 2015) taking into account the following aspects. First, in relation to the need for a major presence of geographic thinking in school teaching, it needs to be incorporated into various levels of the education system. With respect to the weak approach to geographic thinking in the curriculum and the lack of significant geographic themes in teachers' academic preparation, the necessary revisions imply the need for an up-to-date view of geography as a social science. In these new changes, geography includes important themes for students across all levels of education. For instance, included in the new curriculum are: spatial patterns of human occupation of territory; interrelations between society and nature; the holistic conception of the planet as mankind's home; spatial impacts of the globalization process, global warming, migratory flows, rapid urbanization, the location of transnational companies, relations between Chile and its region and the global economy, and foreign treaties and their impact on geographic space. Most importantly, an emphasis on geographic themes is organized according to the concept of sustainable development and the promotion of citizens' geographic knowledge. As such, these revisions leverage the power of geographically thinking, articulating with the essence of GeoCapabilities as:

> Without geography education young people would not be able to understand or even question their position in the world. This does not refer to the knowledge of their hometown's latitude and longitude, but to the understanding of influencing factors between the environment and the individual. Without this, the student would be denied the opportunity of seeing things beyond his own world (GeoCapabilities 2015).

---

[1]This work heavily referenced the National Geography Standards (2012) from the United States.

Geography education in schools has a considerable potential to develop and consolidate the principles of sustainability. The spatial perspective—what characterizes geography as a social science—unites traditional aspects of sustainable development with specific geographic spaces. There is a very close relationship and complementarity between geographic education and citizen formation (Lidstone and Williams 2006). Students' familiarization with the territorial structure of a country and their link with related democratic institutions correspond with an important school experience, fostering their formation as citizens.

In relation to the lack of geographic abilities which are important to the formation of geographic thinking, the curricular revisions consider the following sequence of criteria: first is the concept of spatial location and place, which implies a progression of geographic abilities from first to fifth grade in elementary school. Conceptually, this refers to the knowledge of spatial location and distribution of natural, social, economic, political and cultural processes on the earth's surface. At first these abilities promote a general geographic view of Earth, but they eventually progress to build a more detailed knowledge of the planet, helping students to identify different geographic regions. Second is the concept of interdependence to become aware that human beings transform geographic space and how this, in turn, has an influence on people. This ability is mostly developed between seventh grade and twelfth grade. Through this ability the way that geographic space is seen can be transformed into a dynamic vision which contributes to understanding its richness, diversity and complexity. Finally, there is the systematic understanding of geographic space which refers to the capacity of explaining the spatial dynamic of a specific territory consisting of different geographic dimensions such as proximity and distance. This ability is also mostly developed at the secondary school level. It starts with the establishment of simple relationships between elements, and then becomes a systematic vision that incorporates different variables and allows the linkage of locations to other places on the planet. In total, geographic education has been constantly and explicitly incorporated into the new curricular changes as it is important that new generations develop geographic abilities and geographic thinking in order to be able to analyse and understand territory as a human construct. With these revisions, geography plays a central role in the integral formation of 21st Century citizens in Chile.

8.4 How Teachers Support Geographical Thinking

Our globalized postmodern society has created a profound transformation of the traditional concepts of space and time in geographic education. This is expressed in the tremendous technological development in media communications and the information era, generating a quick and instantaneous proximity among people and worldwide communities with the use of new virtual information technologies such as social media.

Other issues in our global world have geographic-territorial expressions related to this transformation. One example can be seen in the regional and local economic structures associated with a free market economic model, often seen clearly in poorer and emerging regions of the world through rapid urban development, pollution, space segregation and congestion exacerbated by the high rates of income disparities and social inequality. Other examples include changes in ecosystems and the natural environment, expressed in the destruction of natural ecosystems and native species, as well as the pressure of searching for cleaner sources of energy. Furthermore, climate change at the local, regional and global scale may be profound, including increasing desertification in vast swaths of our planet, threats to water reserves due to the constant retreat of glaciers and melting, and increasing sea levels and ocean temperatures. Concurrent with all of these, is the progressive increase of human migrants both inside and between countries (Gilbert M. Grosvenor Center for Geographic Education 2012).

In the dynamic and hectic context of a global world, the educational role of geography as a social science within the school curriculum is strengthened. It explains new relations among societies and the physical world, promoting specialized and rational geographical thinking to tackle social and environmental issues affecting human beings and their bond with nature. Geography education contributes to the acquisition of geographic thinking and geographically well-informed people, thus supporting the idea that individuals with a solid background in geographical knowledge will have greater capabilities overall (GeoCapabilities 2015).

This perspective emphasizes the relevance of geographic education in order to create civic responsibility that benefits children and young people by developing their cognitive skills in geographical thinking, directed towards the systemic comprehension of the contemporary world programmes that have a spatial and geographic aspect. Therefore, new generations can gain geographical literacy, a sense of belonging, empathy before adversity, and an ability to promote agreement, values, and active responsibility.

Developing geographic thinking skills for new generations is a great challenge and an enormous opportunity for school systems and their teachers to reconsider and re-evaluate their educational practice—including their classroom planning and activities, methodological and didactic strategies, resources, and ways to attain cognitive skills. It also provides schools and teachers within the opportunity to acquire new basic concepts and current topics such as: globalization, cultural diversity geospatial technologies, localization, climate change, energy, national and international security, environmental disasters, infrastructure, resource management, and employment. In the classroom environment, integrating the analysis and comprehension of these new concepts and geographic topics has challenged geography education in all of its aspects: epistemological, methodological and didactic. To develop children and young people's abilities and attitudes, there is a need to integrate new resources and technological support such as mobile devices platforms, video conferences and educational and information technologies.

The concern for geography education in Chile dates back to the nineteenth century when the government in the Republican Period decided to create suitable institutions for developing geography education in schools and universities and for promoting studies to describe the whole country. Towards the end of the 1980s, geographical education was taught in a descriptive way and started to change in the 1990s. This is due to, among other factors, the process of professionalizing teachers and the changes experienced in the initial geography teacher training programme in Chilean universities.

Geography teacher training in Chile has been developed mainly in universities and professional institutes. Full-time geography teachers are not provided for first and second grade in elementary education. By contrast, at secondary levels, a qualified teacher is responsible for history, geography and social sciences classes and their teaching is oriented towards the explanation and analysis of the time-space dimensions of societies. Geography education is permanently and explicitly present in the new curriculum arrangement. As a consequence, new generations of students will acquire geographic skills enabling them to understand territory as a human construction characterized by constant change.

Geographic education focuses on sustainability and creating good citizens. These two extremely important trends are evident throughout primary and secondary education. The first trend, geographic education for sustainability, responds to a demand that has profound social, economic and environmental linkages including globalization, climate change, human development, and biodiversity that are important concepts representing both the ethical and stewardship aspects currently taught to new generations. As suggested by the GeoCapabilities project, geographic education for citizenship empowers students to be responsible members of society who can adapt themselves to local environments and develop productive social relationships, including seeking positive change.

In the United States, geography education has evolved considerably over the decades. Throughout most of the 20th Century, geography teachers focused on facts and memorization rather than geographical thinking. Concurrent with this was a geography curriculum that concentrated on area studies such as knowing dimensions and characteristics of world regions. One of the first foundational frameworks for geographic education that indeed centred on geographical thinking was the *Five Themes of Geography* that included location, place, human/environment interaction, movement, and regions, each engaging students in a more thoughtful, systematic analysis of the world around them. Next, the *Geography for Life: National Geography Standards* in 1994 established a set of six essential elements and eighteen standards that dramatically changed K-12 geography education and advanced geographic thinking for both elementary and secondary students. These standards are systematic in scope covering the breadth of the discipline, as well as engaging students in applying geography to the past, present, and future with a focus on problem solving. In 2012, these standards were updated to include more of a focus on contemporary technological aspects of geography such as GIS and remote sensing. Even with this robust framework in place, geographical thinking has progressed even more in recent years with the integration of both literacy and STEM (science, technology, engineering, and mathematics). As an example, the National Geographic Society (2013) created *Common Core English Language Arts and Geography Connections* in 2013, a detailed alignment guide that integrates geographical thinking with literacy. Based on the Common Core English/Language Arts standards that have been adopted by the majority of states and the revised national geography standards, the document supports students in linking geography with reading, writing, speaking, and conducting research. Additional advances in geographical thinking relating to K-12 teaching and curriculum include the *Next Generation Science Standards* (2014) which are now being phased in across the country and include a focus on natural resource management and developing science based solutions for such issues as well as the new *College, Career, and Civic Life (C3) Framework*, that is designed to guide emphasis on inquiry as well as motivating students to actively participate in problem solving (NCSS 2013).

8.5 The Bi-national Project

This project developed out of collaboration that began with a 2009 Fulbright-Hays Group Projects Abroad grant awarded to the Geographic Alliance of Iowa, where a Project Director led twelve K-12 educators on a four week curriculum development project to Chile. The theme of the project was population change and the goal was to use first-hand experience to create standards-based lessons relating to this topic. Inherent to the endeavour was connecting population changes in Chile to those in the participants' home state of Iowa. Common themes included urbanization, rural depopulation, and international immigration. Through this project, a connection was made with Chilean geography education colleagues and thus began a project that focused on sustainability and its commonalities between two far removed regions of the world.

The main objective of this current research was to design and develop bilingual curricular materials for classrooms, based on National Geography Standards (Gallagher and Downs 2012) and Chile's Fundamental Objectives of Geography Learning, from a perspective of sustainable development and geographical thinking. The research was a bi-national effort between Chile and the US, recognizing the key international importance of sustainability and the need to connect local issues, events, and problems with the larger global context and with similar occurrences elsewhere in the world. This is an important outcome in geographical thinking, investigating, gathering information during fieldwork and interpreting, analysing, and developing educational resources. The lesson plans and classroom activities were designed to work with digital media, and were focused on addressing local and global problems related to geographical sustainability and key aspects of geographical thinking, especially interdependence and scale.

Table 8.1 Sustainability themes and focus area by country

General theme	Lesson plan focus for Coquimbo, Chile	Lesson plan focus for Iowa, USA
Economic—energy	Energy challenges in the region	Wind energy
Economic—industrialization	Growth of the Pisco industry	"Green" manufacturing
Environmental—hazards	Tsunami hazards in Coquimbo Bay	Flood hazards in Iowa
Environmental—pollution	Pollution in La Herradura Bay	Water pollution in Iowa
Social—poverty	Rural poverty in La Higuera	Rural poverty in Iowa
Social—urbanization	Urban growth in La Serena-Coquimbo	Des Moines urban change in comparison to Iowa
Cultural—immigration	Italian immigration to the region	Immigration in Iowa
Cultural—agricultural	Water usage in rural Coquimbo	Alternative agriculture

Source Oberle et al. (2015)

This bi-national project enlisted Chilean and North American pre-service geography/social studies teachers to design a series of sustainability lesson plans, with both nations identifying commonalities among the two regions that served as the focus of the project. Essential to these lesson plans is the idea that regional sustainability issues in two different nations not only have much in common with each other, but more importantly represent local examples of global challenges relating to sustainability. The project team identified eight aspects of sustainability that both represent a wide cross-section of that focus area as well as being evident in both the region in Chile and the region in the United States (Table 8.1).

As an example from the United States, the pollution lesson plan focuses on Iowa and was selected because surface water pollution is a serious issue in the central United States as well as having an equivalent in Chile where coastal water pollution presents an environmental challenge. Each of the lesson plans not only include the lesson itself, but also a web quest where students are guided through a process of extending the state or regional based lesson to a wider national example. For the Iowa water pollution lesson, the web quest includes several components. First, students worked on a collaborative project where they were assigned a particular type of water pollution, such as ground water or oceans. The web quest includes resources to assist students in finding information and required the production of visual resources that accompany the narrative. The resources typically include links to government, scientific, non-profit, or other reputable sites, as well as short video clips from YouTube or similar sources. Central to the web quest is a detailed rubric that provides an overview of expectations. In this case, categories include poster aesthetics as well as substantive content, workmanship and collaboration. Figure 8.1 Illustrates the front page of two web quests focused on the same theme of hazards: flooding in Iowa and tsunamis in Chile.

Fig. 8.1 Front page overview of web quests that focus on the common theme of hazards in each of the two countries

8.6 Geographical Thinking About Sustainability

Sustainability is an ideal means for linking physical and human geography, especially in terms of creating a curriculum that supports geographic thinking. The National Council for the Social Studies (NCSS), in coordination with the US Partnership for Education for Sustainable Development, published a draft of standards to promote "sustainability literacy" (US Partnership 2009). These standards include grade bands from the K-12 level and address a range of geographic topics such as systems thinking, the depletion of shared resources, environmental justice, urban design, global health, and multilateral organizations (US Partnership 2009). In terms of geo-literacy, geographic perspectives, and geographic thinking, sustainability is central to each. The theme is inherent in geo-literacy in both interactions and interconnections and the many decisions about sustainability are inexorably linked with implications. Sustainability is a model topic for geographic perspectives too, particularly to demonstrate interdependencies among scales where local decisions have global ramifications and global issues quickly affect local communities. In the realm of geographic thinking, sustainability emerges in all four concepts, particularly *scale and connection* and *proximity and distance*.

Lambert (2004) emphasises the importance of geography in integrating human and physical aspects and, by extension, provides an example for how sustainability could serve as that link. He states:

> For example, a very special aspect of geographical study is its refusal to separate the physical and human worlds. Thus, geography is the subject resource that allows children to discover the importance of 'holistic' decision making - where to put that building, for example, or what benefits may follow the planting of trees along this street' (Lambert 2004 p. 1).

8.7 Developing Students' Geographical Thinking About Sustainability

This section discusses the role physical and human geography play in developing geographical thinking in geographic education for sustainability. These topics describe the sequence of learning development in certain areas (or domains) considered essential for the education of students at each level. The curriculum for Geography and Social Sciences has the purpose of developing students' knowledge, skills and attitudes. This allows them to organize their understanding of our world, either through its physical and human geography, and prepares them for responsible stewardship and citizenship. It is expected that students will be able to comprehend the connections between society and the natural environment and appreciate the significance of environmental balance. This element of knowledge, coupled with action, articulates with Geocapabilities and its intent to think and act from different perspectives (Geocapabilities 2015).

Physical and human geography centre geographic thinking around three interrelated aspects, with the first being spatial location and the systematic comprehension of geographic space. This refers to the knowledge of location, the spatial distribution of elements and geographic processes, and the comprehension of spatial dynamics of a particular territory. It incorporates several variables including natural, social, economic, political and cultural aspects. This concept starts from a general geographic vision of Earth and the identification of simple connections among geographic elements, and then progresses toward a more detailed and systematic knowledge of the planet and an understanding of the interrelations between different variables of geographic space. The next aspect is the ability to analyse geographic space. This refers to the development of skills related to the direct observation and interpretation of geographic space, applying geographical categories of increasing complexity in order to analyse significant geographical problems, and formulate hypotheses about their causes and impacts. The last aspect is valuing and possessing a responsible attitude about geographic space. This entails the development of attitudes of stewardship and both a personal and collective responsibility for geographic space, realizing its role in environmental sustainability and in valuing a sense of place, not just of one's town, but of the whole planet. In the curriculum there is the assumption that territorial order is a human construct, which can be modified to benefit the greater good and quality of life.

Since the theme of sustainability invariably links physical and human geography, we present two project lesson plans that demonstrate, in detail, how this develops geographical thinking. We employ Peter Jackson's geographical thinking concepts to guide this outline of the lesson plans and associated web quests. While all sustainability topics integrate aspects of both physical and human geography, the joint lesson plans on hazards are particularly effective for highlighting this connection. Both Iowa and the Coquimbo region of Chile are at high risk of natural disasters, flooding in the Iowa context and tsunamis in the case of Coquimbo and all of coastal Chile. The need for sustainability is clearly evident in both lesson plans, with agricultural practices and urbanization in Iowa contributing to an enhanced flood risk and sea level rise due to global warming intensifying the tsunami risk in coastal areas. The Iowa lesson plan *Iowa floods of 2008* engages students in a geographic analysis of flood risk in the state followed by a collaborative research project where students represent a hypothetical committee to develop a proposal to reduce future flood risk. The Chilean counterpart to this lesson plan, *Tsunami risk in Coquimbo Bay*, provides background information about tsunamis, explains the difference between risk and vulnerability, and then asks students to identify areas in the local community that are at the greatest risk of experiencing a tsunami. The web quests associated with this lesson plan include expanding the scale of the research to the entire United States for flooding and, likewise, extending tsunami research to the whole of Chile and other parts of the globe.

Three of Jackson's (2006) four concepts drove this project and demonstrated how it advances geographical thinking. With regards to *space and place*, each of the two focus areas in the lesson plan, Iowa and Coquimbo, are unique places, but their uniqueness is also connected to other more distant places. Flood risk in Iowa is part of its location in between two great rivers, but also is linked to its cultural heritage where generations of Iowans have modified its river systems with the firm belief that human ingenuity will ultimately mitigate flood hazards. Yet, Iowa is also inexorably linked to upstream rain events, while state-wide policy about rebuilding after floods is directly connected to federal policy. Similarly, Coquimbo's coastal population centres and sparsely populated interior are part of its place characteristics, linked to maritime traditions and the economy of its port and fishing industry, but also to the Pan-American Highway and its key importance as the link in a country with an elongated north and south geography. *Scale and connections* are inherent to these lesson plans. In Iowa, scale is evident in flood policy where severe flood events, such as the Iowa floods of 2008, raise national awareness about flood policy including questions about rebuilding and response, while at the same time Federal Flood Insurance and related national policies have a direct and immediate local impact for those Iowa communities along rivers and flood plains. Tsunami awareness and preparedness in Coquimbo is affected by changing national policies resulting from the devastating Chilean tsunami in 2010 and the global response to the Japanese tsunami in 2011. In the case of each of the lesson plans, students are asked to analyse the hazard well beyond just their home state or local region, thus reinforcing the perspective of *scale and connections*. *Proximity and distance* is a common thread in terms of the traditional sense of measured distance such as evacuation routes for tsunamis and distance from flood plains. However, the notion of socially-constructed distance is evident too. In fact, Jackson (2006) discusses the response to the 2005 Asian tsunami as his example of how in our contemporary globalized world, a regional event can galvanize a world community to action. This translates directly to Iowa floods or tsunami risk in Coquimbo (recently, in September 2015, Coquimbo was affected from the devastating local tsunami) where something in a far off state, region or a distant part of the globe can not only raise awareness but also prompt action and promote positive change.

The idea of GeoCapabilities is inherent in these specific examples and broadly across the project. Both tsunamis and flooding are local risks that have an immediate impact on the students in their respective regions. These are also actionable issues where students can indeed develop the capabilities to have an impact and make a change, perhaps educating coastal residents about the best evacuation routes for a tsunami or encouraging inland citizens to be more aware of flood risk and to take proper precautions. These issues are national and global as well since both of the respective countries strive to develop policies and plans to mitigate damage and loss of life from these disasters, as well as the likelihood that a changing global climate would exacerbate the risk for and damage from both types of disasters.

8.8 Conclusions and Recommendations for Future Research

This chapter develops geographic thinking by employing sustainability as a means of incorporating physical and human geography. We developed a series of standards-based lesson plans and associated curricular materials on the theme of sustainability that link the Coquimbo region of Chile with the state of Iowa in the United States. Geography plays a profound role since it masters how to deal with important relationships between humans and the territory they inhabit, both in rural and urban environments. This significance has been fuelled by the concern for the global environment which has a particular impact on the United States and Chile.

Geographic thinking encompasses perspectives, skills, and concepts that are needed to analyse and explain our world. Since this is a bi-national project, we describe geographic thinking in the Chilean context, comparing it to that in the US.

This perspective emphasizes the relevance of geographic education in order to create civic responsibility that benefits children and young people by developing their cognitive skills in geographical thinking, directed towards the systemic comprehension of the contemporary world programmes that have a spatial and geographic aspect. Therefore, new generations can gain geographical literacy, a sense of belonging, empathy before adversity, and an ability to promote agreement, values, and active responsibility.

We value the contributions of teachers and their role in strengthening geographical education. It is important to emphasize that such concepts as sustainability, geographical thinking, and new information and communication technologies encourage greater understanding of space and geography. This approach allows citizens to be integrated with a globalized world in a positive and productive fashion. All this action is accomplished by developing spatial abilities that are honed to achieve a sustainable vision of natural and human resources. This philosophy, embedded in the notion of human capabilities, embodies the ideal of GeoCapabilities and the potential for people to employ more robust geographical thinking to understand their role in the world and act upon that information to make positive decisions that support the greater good.

The challenges and directions for future research about geographical thinking and sustainability includes developing geographic thinking skills for new generations—both a great challenge and an enormous opportunity for school systems and their teachers as they reconsider and re-evaluate their educational practice, including their classroom planning and activities, methodological and didactic strategies, resources, and means to attain cognitive skills. It also provides schools and teachers with the opportunity to acquire new basic concepts and current topics such as: globalization, cultural diversity, geospatial technologies, localization, climate change, energy, national and international security, environmental disasters, infrastructure, resource management, and employment. In the classroom environment, integrating the analysis and comprehension of these new concepts and geographic topics has challenged geography education in all of its aspects: epistemological, methodological and didactic.

The research should deepen a bi-national effort between Chile and the US, recognizing the key international importance of sustainability and the need to connect local issues, events, and problems with the larger global context and with similar occurrences elsewhere in the world. We highlight the need to develop multinational research in the field of geographic education, allowing the exchange of methodological experiences, educational strategies and development of learning resources between different realities, targeted to development of geographic thinking to encourage the new generations with a dynamic understanding of space, integrating a multitude of natural, social, economic and political variables (Brooks 2015). Geography helps people to think critically about sustainable living locally and globally and how to act accordingly. This is an important objective in geographical thinking, investigating, gathering information during fieldwork and interpreting, analysing, and developing educational resources. Through the perspective of sustainability, there is the great potential for the development of citizenship, such as attitudes of care and responsibility to the environment in which one lives, thus realizing that achieving sustainable development is a challenge that concerns us all - society, the state, and as individuals.

Acknowledgements Dr. Araya is grateful for support of a grant from DIULS (# PR14412), Research Direction University of La Serena.

References

Butt, G. (Ed.). (2011). *Geography, education and the future*. London: Continuum International Publishing group.

Brooks, C. (2015). International differences in thinking geographically, and why it matters. In *The power of geographical thinking, Institute of Education*. University of London.

Edelson, D. (2011). *Geo-Literacy. Preparation for far-reaching decisions*. National Geographic, Education Online Resources (NatGeoEd. com), National Geographic Society.

Environmental Protection Agency, EPA. (2014). *Sustainability*. http://www.epa.gov/sustainability/basicinfo.htm#sustainabilityLast. Accessed June 25, 2015.

Gallagher, S., & Downs, R. (Eds). (2012). Geography for life: National geography standards. In *Geography education national implementation project (GENIP)*. Washington: National Council for Geographic Education.

GeoCapabilities. (2015). Last accessed December 20, 2015. http://www.geocapabilities.org/

Geographical Association. (2009). *A different view: A manifesto from the geographical association*. Sheffield, UK: Geographical Association.

Gilbert M. Grosvenor Center for Geographic Education. (2012). Why geography is important. National Council for Geographic Education; National Geographic Society; Association of American Geographers. EE.UU.

Jackson, P. (2006). Thinking geographically. Geography *91*(3), 199–204 (Conference in geographical association). [On line] http://www.geography.org.uk/projects/e-scape/thinking geographically/#top. Last accessed June 25, 2015.

Jackson, P. (2009). *"Thinking geographically" in a different view: a manifesto from the geographical association*. London, UK: Geographical Association.

Lambert, D. (2004). *The power of geography*. London, UK: Geographical Association.

Lidstone, J., & Williams, M. (Eds). (2006). Geographical education in a changing world. In *Past experience, current trends and future challenges. The GeoJournal Library* (Vol. 85). The Netherlands: Springer.

National Council for the Social Studies. (2013). *College, career, and civic life (C3) framework*. Silver Spring, MD: National Council for the Social Studies.

National Geography Standards. (2012). *Geography for life*. Washington DC: National Geographic Society.

National Geographic Society. (2013). *Common core english language arts and geography connections*. Washington DC: National Geographic Society.

National Research Council. (1997). *Rediscovering geography: New relevance for science and society*. Washington DC: National Academy Press.

Next Generation Science Standards. (2014). *Next generation science standards*. Washington DC: Achieve.

Oberle, A., Araya, F., Cortés, X., & Ullestad, M. (2015). Developing a bi-national geography curriculum in sustainability. *The Geography Teacher, 12*(3), 108–117.

US Partnership for Education for Sustainable Development. (2009). *National education for sustainability K-12 student learning standards*. Washington DC: US Partnership.

Chapter 9
Using Weblogs to Determine the Levels of Student Reflection in Global Education

Nina Brendel

9.1 Introduction

In the recent discourse on geographical thinking, much attention and thought has been given to the teacher on the one hand (Catling 2015; Fögele 2015; Seow 2015) and the role of the curriculum on the other (Mitchell 2015; Pauw and Beneker 2015; Young 2013). However, this study particularly focuses on high school students and how they reflect on geographical content in everyday classroom situations.

Reflecting on geographical issues and processes is an essential part of geographical thinking, especially when dealing with complex topics of global education such as sustainability or globalization. Students' reflections on global processes have gained great importance in recent years, most notably in the light of the rise of Education for Sustainable Development (ESD). In order to 'take informed decisions and responsible actions', educators must 'rethink learning environments' to promote 'core competences, such as critical and systemic thinking' (UNESCO 2014, p. 12). Global Education—a concept closely linked to ESD—similarly 'focuses on supporting active learning and encouraging reflection with active participation of learners and educator' (O'Loughlin and Wegimont 2002, p. 149). Due to the general character of these international documents, practical advice on how educators can promote *critical thinking* and *encourage reflection* is not usually given. These general targets are concretized in national documents, such as the German 'Orientierungsrahmen für den Lernbereich globale Entwicklung' (or, 'framework for the educational field of global development'; Appelt and Siege 2008). Here

N. Brendel (✉)
University of Muenster, Muenster, Germany
e-mail: nina.brendel@uni-muenster.de

C. Brooks et al. (eds.), *The Power of Geographical Thinking*,
International Perspectives on Geographical Education,
DOI 10.1007/978-3-319-49986-4_9

Global Education aims to promote three basic competences (Appelt and Siege 2008, p. 72):

(a) *Competences to think in systems*, which imply constructing knowledge, comprehending complex interrelations and finally a holistic understanding of systems
(b) *Competences of assessment*, which base evaluations on critical reflection of content and values
(c) *Competences of action*, which enable students to take responsibility for their actions in accordance with the objectives of sustainable development.

While reflection is directly referred to only within the competences of assessment, thinking about geographical content is also essential in order to comprehend interrelations between facts and to grasp the nature of systems (competences to think in systems). Similarly, students need to have reflected on situations or issues before they can take informed actions (competences of action). It is therefore argued that processes of reflection are an essential part of *all three* competences of Global Education, as defined by the German framework.

However, to distinguish and determine students' reflective thinking is often a challenge for educators in geography lessons. In order to provide teachers with a tool to recognize and foster students' reflections, the study described here aimed to distinguish levels of reflective thinking amongst students and determine factors that foster or decrease student reflection in geography. The findings and their implications shed light on how student reflection can be fostered to promote powerful geographical thinking.

9.2 Theoretical Models on Reflective Thinking and Reflective Practice

Despite their prominence, the aforementioned documents neither define what is meant by *reflection* nor explain how it can be evaluated or categorised by educators or students. At this point, it is important to take a closer look at what is meant by *reflection* and how it is linked to *thinking about geographical content.*

Much has been written about the concept of *reflection*, often such work is based on the thinking of John Dewey. As early as 1910 Dewey defined his concept of *reflective thinking*, which seems an appropriate term for reflecting on—in our case geographical—content:

> *Active, persistent, and careful consideration of any belief or supposed form of knowledge in the light of the grounds that support it, and the further conclusions to which it tends,* constitutes reflective thought (Dewey 1997, p. 6, original emphasis).

The important thing to notice here is that Dewey is referring to thinking about *content*. Transferred to an educational and geographical context, this implies that students ponder topics such as globalization, climate change and sustainability, and draw conclusions from this.

In the course of the 20th century, however, research on reflection rather shifted its focus from thinking about *content* (reflective thinking) to thinking about *actions* (reflective practice). Building on the ideas of Dewey, Schön (1983) distinguishes two types of *reflective practice: reflection-in-action*, the consideration of someone's behaviour while acting in a professional situation, and *reflection-on-action*, which denotes analysing one's actions, given alternatives and consequences in retrospect (Schön 1983). It is important to note that Schön does not only refer to teachers, but also understands that students can also be reflective practitioners (Schön 1983, p. 52).

Other popular models of reflective practice were presented by Gibbs (1988) and Kolb and Fry (1975), both of which are often used to prompt and/or analyse reflection (Harrison et al. 2003; Healey and Jenkins 2007; Konak et al. 2014; Ozdemir 2015; Paterson and Chapman 2013; Su 2014; Timmins et al. 2013). Due to being easily comprehensible, these two models seem to be well-suited to equip students with a basic framework for written reflection. The concept of *reflective thinking,* as described by Dewey, is most appropriate when investigating the levels of reflection of students' thoughts about geographical issues due to its focus on *content* rather than *actions*.

The theoretical framework being set, there still remains the question of how to 'measure' students' reflective thinking. In an effort to distinguish different qualities or stages of reflection, diverse stage models have been developed on an empirical basis (Zeichner and Liston 1985; van Manen 1977; Kember et al. 1999; Hatton and Smith 1995). Most of these models understand reflection in the sense of *reflective practice*, meaning reflection on *actions*. If educators aim to determine the level of students *reflective thinking* (i.e. thinking about *content*), far fewer models have been developed. A widely acclaimed stage model to categorise reflective thinking is proposed by Bain et al. (1999). Their model served as a framework for many other empirical works (Carrington and Selva 2010; Chen et al. 2009; Henderson et al. 2004; Hsieh et al. 2011). Chen et al. (2009), for instance, based the coding themes for their study on the five-level-scale used by Bain et al. Table 9.1 shows both models in comparison.

In the study reported here, the model of Bain et al. (1999) and its adaption by Chen et al. (2009) was used for three reasons: firstly, the models have been used extensively in prior research and are considered to serve as a reliable tool to determine the levels of reflection. Secondly, apart from very few exceptions, the models only refer to *reflective thinking* as defined by Dewey, not *reflective practice*. Thirdly, they provide clear criteria for each level of reflection which allows for a transparent grading of reflective thinking.

Table 9.1 Reflection scale by Bain et al. (1999, p. 60) and revised by Chen et al. (2009, p. 286)

Levels of reflection	Five-point level of reflection scale by Bain et al. (1999)	Evaluation criteria by Chen et al. (2009)
Level 1 reporting	The student describes, reports or re-tells with minimal transformation, no added observations or insights	Only repeated the content already in article
Level 2 responding	The student uses the source data in some way, but with little transformation or conceptualisation The student makes an observation or judgement without making any further inferences or detailing the reasons for the judgement The student asks a 'rhetorical' question without attempting to answer it or consider alternatives The student reports a feeling such as relief, anxiety, happiness, etc.	Just used few concepts Stated observed phenomena, no reasoning Stated personal affection only
Level 3 relating	The student identifies aspects of the data which have personal meaning or which connect with their prior or current experience The student seeks a superficial understanding of relationships The student identifies something they are good at, something that they need to improve, a mistake they have made, or an area in which they have learned from their practical experience The student gives a superficial explanation of the reason why something has happened or identifies something they need or plan to do or change	1. Stated relationships within the context 2. Explained the cause
Level 4 reasoning	The student integrates the data into an appropriate relationship, e.g. with theoretical concepts, personal experience, involving a high level of transformation and conceptualisation The student seeks a deep understanding of why something has happened The student explores or analyses a concept, event or experience, asks questions and looks for answers, considers alternatives, speculates or hypothesises about why something is happening The student attempts to explain their own or others' behaviour or feelings using their own insight, inferences, experiences or previous learning, with some depth of understanding The student explores the relationship between theory and practice in some depth	1. Detailed explanation of the rationale 2. Combined theory and practice

(continued)

Table 9.1 (continued)

Levels of reflection	Five-point level of reflection scale by Bain et al. (1999)	Evaluation criteria by Chen et al. (2009)
Level 5 reconstructing	The student displays a high level of abstract thinking to generalise and/or apply learning The student draws an original conclusion from their reflections, generalises from their experience, extracts general principles, formulates a personal theory of teaching or takes a position on an issue The student extracts and internalises the personal significance of their learning and/or plans their own further learning on the basis of their reflections	1. Expressed high order of reconstructing 2. Organized theory, rule and experience in a systematic approach

9.3 Weblogs Used for Reflective Journal Writing

In order to determine the level of student reflection on geographical content, personal weblogs written by each student for every geography lesson during a time period of four to eight weeks were used (depending on the length of the lesson unit).

While student reflection was formerly assessed by the use of hand-written journals, recent studies often use online tools such as weblogs or online portfolios to evaluate student reflection (Andrusyszyn and Davie 1997; Mansvelder-Longayroux et al. 2007; Oner and Adadan 2011; Sharma 2008). Journal writing has proven useful to enhance reflection in teacher education (El-Dib 2007; Hramiak et al. 2009; Morrison 1996; Naghdipour and Emeagwali 2013; Wopereis et al. 2010; Yang 2009), as well as in geography in higher education (Chappell 2006; Haigh 2001; Park 2003). While blogging is extensively used in higher education, weblogs also seem to be a promising tool in secondary education (Huffaker 2004; Hung 2014; Overby 2009; Richardson 2010; West 2008)—and for good reasons, as digital journal writing enhances the communication with the instructor and leads to a personal construction of knowledge (Andrusyszyn and Davie 1997). Students are very familiar with these new technologies from their everyday lives, and so they have great potential for the classroom (Overby 2009). Moreover, competences regarding digital media that students acquire through informal learning are highly valued and motivating for formal learning in schools (Dabbagh and Kitsantas 2011). While traditional learning journals were solely written for, and read by, the teacher, digital journals (e.g. weblogs) make it very easy to publish students' work beyond the classroom. In this way, students can share what they have worked on with their parents, friends or family. Furthermore, students can connect with people working on similar projects via social media and eventually contribute to public discourse in a community of practice (Yang 2009). However, an essential prerequisite for all online activities is that teachers heighten the students' awareness regarding copyright issues (e.g. Creative Commons) and also how to act safely online.

9.4 Study Design and Methods

As discussed above, weblogs seem to be a suitable tool for student reflection. However, previous studies primarily focused on college or university students on the one hand and on *reflective practice* rather than *reflective thinking* on the other. In the study presented here, the question was raised about how intensively high school students (aged 15–16) think about the geographical content discussed in geography lessons (Table 9.2). Therefore, students on four geography courses (grade 10 and 11, German secondary school) were asked to reflect on the lessons' content in a personal weblog during a four to eight week unit on global education.[1] These blogs were protected by password and could only be accessed by the students of this class, their teachers and the researcher. Additionally, the students were asked to collaboratively write a wiki in order to link each lesson's contents. A final blog entry at the very end of the unit offered the chance for the students to give feedback on the project as well as on blogging in school in general.

While the main interest of the study was to determine students' reflective thinking on geographical content, teachers were also asked to write about their lessons, the decisions they made and their (re)actions in class. This data was included in the project, because it was assumed there might be a connection between the reflective practice of teachers and their students' reflective thinking skills. In addition, qualitative, semi-structured interviews were conducted with all teachers beforehand in order to understand and document background information, such as the learning setting or the relationship between the teacher and their class from the teachers' point of view. Just like their students, the teachers could also evaluate the project in a final blog entry.

In order to determine the levels of reflective thinking achieved during the unit, an inductive-deductive content analysis of the student weblogs was conducted based on the five-level reflection scale of Bain et al. (1999) and its revision by Chen et al. (2009). On the one hand, each sentence of each weblog article was coded with the corresponding level of reflection based on a fusion of the criteria given in both models. On the other, the reflection scale based on the findings in our data was adapted. In this circular manner, a multi-level reflection model for global education was developed, based on existing and empirically proven reflection scales as well as on the original data of this study. In order to determine factors which influence reflective thinking, teachers' weblogs, the teacher interviews, the wikis and the evaluations were analysed using inductive and deductive content analyses

[1]The topics of the unit comprised sustainability, sustainable tourism development and water resources.

Table 9.2 Study design

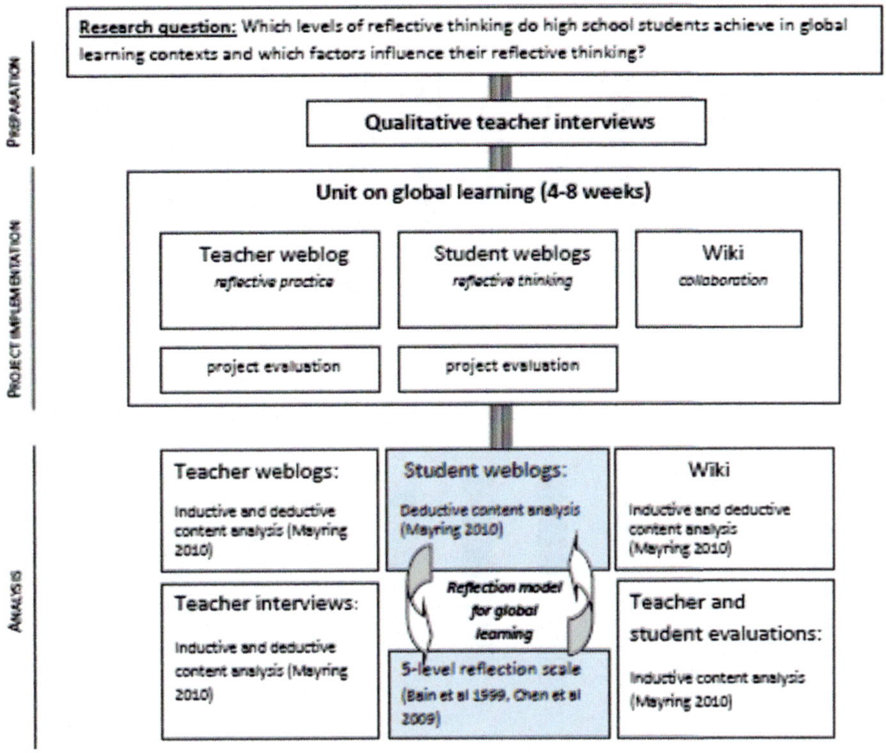

suggested by Mayring (2000). Bringing all findings together, strategies to support individual reflective skills of high school students were presented to help equip educators with the methods to differentiate with regard to reflective thinking.

9.5 Selected Findings

After all data had been analysed and interpreted significant differences were found in the reflective thinking among students of the same class, a variation of reflective thinking performance between different lessons, as well as large differences between the classes participating in this study. Figure 9.1 shows the levels of reflective thinking achieved by all students in class A and class C during their units on global education.

With the levels of reflective thinking being determined, factors that account for those differences were sought. The findings discussed here promise to be important for the understanding of geographical reflective thinking.

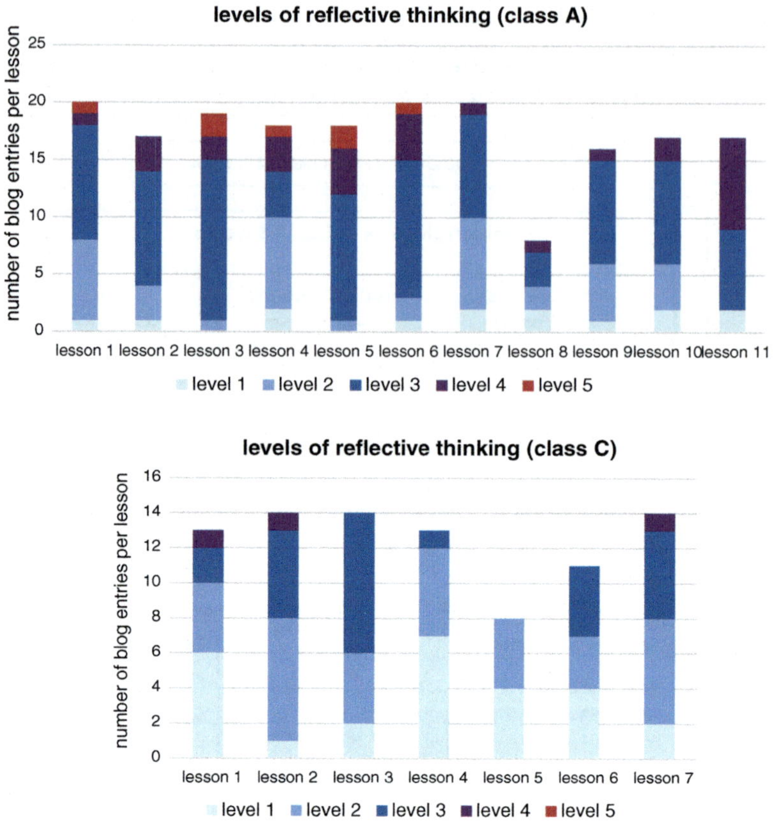

Fig. 9.1 Levels of reflective thinking per lesson in class A and C

9.6 The Power of Everyday Knowledge on Geographic Reflection

The most prominent factor to influence students' reflective thinking was found to be the interconnection of students' everyday life with what was taught in geography lessons.

Two out of the four instructors chose this approach to introduce their class to the topic of the unit, which resulted in increased reflective thinking in the students' weblogs: At the very beginning of the unit on water resources, the teacher of class A asked her students to write down how much water they consumed in one day (lesson 1). During the unit, the students tried to save water as best they could. Again, at the end of the unit, the students kept a diary on their water consumption for one day. Surprisingly for the class, only very few students had actually managed to reduce their water consumption. These results triggered an intensive debate in class as well as in the weblogs and led to the highest levels of reflective thinking achieved by the students of this class (see Fig. 9.1, class A, lesson 11). The increase in reflective thinking in the last *and* the first lessons indicate that the linkage of the lessons' contents, on the one hand, and the students' daily behaviour and experience, on the other, increases student reflection significantly.

This assumption is supported by the case of class B: The teacher of this class used the students' daily habits to make them realize how much every one of them contributes to global carbon emissions: she asked the students to calculate their own carbon footprint, based on their way of living, their nutrition, their mobility and their purchasing habits. Similar to class A, the students' weblogs showed an increase in reflective thinking performance in this lesson due to the connection of theoretical concepts to their everyday lives.

Furthermore, the content analysis of the weblogs in *both* courses revealed that student reflection increased, when personal as well as global consequences of geographical processes or developments were discussed and related to each other. Students who showed rather low reflective thinking abilities in their weblog articles (mostly level 1 or 2), profited especially from this approach and achieved a higher level of reflective thinking (mostly level 3). Moreover, in class C, some students showed their highest level of reflective thinking in a lesson that dealt with a national park they had visited earlier in the year. In their weblogs, they related their personal experiences to the concepts discussed in school and thus achieved high levels of reflective thinking. A lack of reference to the students' everyday experiences, however, resulted in a negative effect of reflective thinking performance: lessons that could hardly be connected to the students' daily life experiences triggered the lowest levels of student reflection (see Fig. 9.1, class C, lessons 4 and 5).

All four participating teachers were either directly or indirectly aware of the effects of everyday experiences on reflective thinking, because all of them mention in some way a relationship between the two aspects in the interviews. Nevertheless, findings from our data show the importance of addressing the effects of individual decisions or behaviour on global processes, as well as their feedback effects. It is important to stress that the interconnection of everyday knowledge drawn from individual experience and what is taught in geography lessons serves best to foster reflective thinking, if clear focus is placed on their reciprocity.

In the recent discourse on powerful knowledge, however, everyday knowledge is regarded as the opposite of scientific knowledge. Referencing his ideas to the distinction of everyday and scientific concepts made my Vygotsky (Young 2008, p. 52), Michael Young defines *powerful knowledge* as knowledge that can only be acquired in school and that is disconnected from everyday knowledge (Young 2011, p. 150):

> It is also, if not always consciously, what parents hope for in making sacrifices to keep their children at school; that they will acquire powerful knowledge that is not available to them at home. Powerful knowledge in modern societies in the sense that I have used the term is, increasingly, specialist knowledge.

Although he concedes that knowledge that students acquire outside school should not be disregarded, he claims that students should not be involved in the selection of knowledge due to their lack of prior knowledge (Young 2011, pp. 150–151). Lambert (2015) agrees that it lies within the responsibility of the instructor and the scientific community to determine what is powerful knowledge. Similarly the Geographical Association's national curriculum review group distinguishes between geographical and everyday thinking:

> geographical thinking is not everyday thinking. Neither are schools everyday places: they introduce children to the world as an 'object of thought' rather than as a 'place of experience'. Subjects help organise this thinking by relating concepts systematically. For example, when pupils learn about 'the city' as an object of thought they are taken beyond the realm of their experience by learning about form and function, or about economic and social processes. They are therefore introduced to 'theoretical' concepts which are systematically related to each other and which require different thought processes from those of everyday learning: for example, they enable us to make links, comparisons and generalisations (Geographical Association's national curriculum review group 2012, p. 7).

Without a doubt, teaching students to think in systems is one of the unique goals of geography education in school (KMK and BMZ 2015, p. 10). However, the findings of this study indicate that an increase in student reflection on geographical issues can be achieved particularly by connecting everyday experiences with theoretical concepts. These results are consistent with the position of Margaret Roberts, who does not entirely agree with Young's ideas and stresses the importance of everyday knowledge in the construction of geographies: from childhood on, people experience geography by where and how they live, what they eat, how they communicate, socialize and so on, which leads to the development of personal geographies (Roberts 2013). Roberts considers the experiences students bring to school as a richness from which geography can draw and as equally important as the content selected by the instructor (ibid). Consequently, she claims:

> I think that if everyday knowledge is important in geography, I think it is in all other subjects, it needs to be made an object of study in the same way as the academic geography (Roberts 2013, Minute 10:02–10:15).

Geography lessons should make students aware of their personal geographies as well as scientific concepts in order to help students realize how geographies interact with their daily lives. This builds on a different interpretation of Vygotsky's concepts of everyday and scientific knowledge than that expressed by Michael Young: whereas Young underlines the separation of both concepts in Vygotsky's work (Young 2008, p. 52), Roberts stresses the mutual influence of both concepts on each other (Roberts 2013). As a consequence, Roberts argues: 'it's not just about developing the school concepts, but school has a responsibility for developing the everyday concepts' (Roberts 2013, minute 05:15–05:35).

As the findings from this study show, the interconnection of everyday experiences and school concepts results in a significant increase in reflective thinking. So, if high levels of reflective thinking are regarded as a part of powerful geographical thinking—in the sense that they empower students to handle geographic knowledge—theoretical geographical concepts should be taught intertwined with everyday experiences in order to promote geographical reflective thinking.

9.7 Key Concepts in Geography to Promote Reflection

As Peter Jackson points out, geography is far more than teaching topographical facts; the great potential of geographical thinking lies in a 'unique way of seeing the world' (Jackson 2006, p. 199), which many researchers and educators have tried to conceptualize in key concepts. An overview of the variety of geographical concepts was summarized by Taylor (2009). Findings from the study confirm Jackson's statement, as the implementation of some specific key concepts into geography lessons proved to be very beneficial in terms of reflective thinking.

A prominent geographical concept that is found in most classifications is that of *scale*. In all four courses, we observed higher levels of reflective thinking in students' weblogs, when geographical content was discussed on a personal or local level as well as with regard to its global implications and interrelations. The important thing to stress here is that those dimensions should not be addressed separately. The focus rather lies on the 'connections between scales', which is also in line with how Jackson (2006, p. 200) understands the concept of scale.

A second concept that promoted reflective thinking in this study is the concept of *physical and human processes*, as described in the 2008 National Curriculum Statutory requirements for key stages 3 and 4 (Qualifications and Curriculum Authority and Department for Children, Schools and Families 2007, p. 103). In German geography, the *human-environment-interactions,* as defined by Weichhart (2003), form the so-called *third column of geography* alongside physical and human geography. The study of such interactions has increasingly gained importance. Human-environment interactions are understood as an independent and autonomous part of geography that focuses on questions, which can only be answered by the combination of physical and human geography (Gebhardt et al. 2007). Integrating both sides systematically, as two sides of the same geographical coin, is key here.

To illustrate this point, let us take a deeper look at the students' weblogs on their geography lessons: in one lesson of the unit in class A, students discussed the local consequences of the global water crisis on their hometown. Apart from addressing different scales, the focus lay on the interrelations of aspects of physical geography (e.g. natural water distribution on earth) and aspects of human geography (e.g. geopolitics). Two weeks later, the students discussed the water cycle including the physical processes of condensation and precipitation. However, aspects of human geography such as the effects of agriculture or industry, for instance, were not mentioned at all. While the weblogs written on the first lessons showed a high performance in reflective thinking by a large part of the class, students' reflective thinking on subsequent lessons was more limited.

A concept closely linked to the aforementioned is that of *environmental interaction and sustainable development* (Qualifications and Curriculum Authority and Department for Children, Schools and Families 2007, p. 103), which is also related to Education for Sustainable Development (ESD). A model that is widely used in German schools to help students understand the dimensions and interrelationships of sustainability is the *sustainability square* (Appelt and Siege 2008, p. 28, Fig. 9.2).

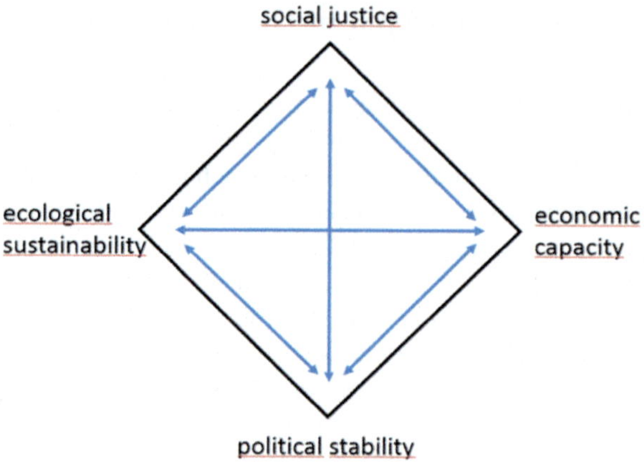

Fig. 9.2 Sustainability square, adapted from Appelt and Siege (2008, p. 28)

The sustainability square was discussed in all four classes, however, the approaches varied significantly. Taking the example of tourism on Bali, the teacher of class B prepared a learning setting that challenged the students to work out the four dimensions of sustainability by themselves. Subsequently, the students discussed whether tourism on Bali was sustainable or not—always taking into consideration all four dimensions of the square and their interrelationships. Although this model was discussed and applied to one example in only one lesson, this inductive approach led to the second-highest amount of higher level reflection (level 4 and 5) of all students in this unit.

Even though class C discussed the sustainability square on several occasions during the unit, their weblog articles showed only low levels of reflective thinking. This might be due to the fact that the class discussed the four dimensions separately (one dimension per lesson) and disregarded the interrelations of all dimensions almost entirely. In this way, students were not able to understand the case study they discussed as a system, and that sustainability relies on a balance of all four dimensions. Consequently, they did not address the concept of *environmental interaction and sustainable development*, despite the fact that they talked about sustainability.

To sum up, the results from the study indicate that the use of geographical key concepts does improve students' reflective thinking considerably.

9.8 The Effects of Gender

Another factor that seems to have an impact on students' reflective thinking is their gender. Due to the qualitative approach of this study the data given here are by no means representative. Statements about the effect of gender are limited to the participants of this study. However, there are intriguing tendencies in the data—which we do not want to disregard. For organizational reasons, data about the students' gender is only available for two of the four courses. Nevertheless, those two classes paint a very similar picture of how reflective thinking can be increased depending on the gender of the students. In both courses, boys achieved higher levels of reflective thinking when they could structure their learning process in an analytical way. For instance, boys profited from the use of the sustainability square by using it as a scheme template, which they applied to various case studies as a guide for their analysis. Moreover, working with a hypothesis (as described above) turned out to be an effective method to increase the male students' reflective thinking.

The girls on the other hand achieved higher levels of reflective thinking when they could work collaboratively. Lessons with a focus on cooperative learning methods (e.g. group activities, learning by teaching) had a positive effect on the reflective thinking girls showed in their weblogs. Furthermore, the girls' reflective performance improved when geographical content was connected to their own experiences and everyday life. While this is a factor that is relevant for both male and female students (as discussed above), girls seem to need this connection more urgently than their male classmates. In one class some female students made an effort to illustrate the individual relevance of each lesson's topic for their personal lives, while the boys' weblog articles did not consider those aspects at all.

In order to determine whether these tendencies are valid in general, quantitative studies with larger groups should be carried out.

9.9 The Relevance of Learning Environments

According to the results of this study, reflective thinking performance also depends on the learning environment. In her weblog, one teacher expresses her amazement at the sophisticated weblog entries of a student, who had rarely shared such thoughts with the class:

> I was really astonished when I found out who the girl behind the nickname [student's nickname] was: a very quiet, but diligently working girl. In my class, she showed only little signs of reflection. But reading her blog made me change my mind about her (excerpt from a teacher's blog).

Apparently, this student was able to show higher levels of reflection, when she could take her time to think about the lessons' topics at home and express her thoughts in writing. Reflective writing in a weblog seemed to be a better-suited learning environment for this student than a discussion with many students in class. Triggered by this experience, the teacher decided to include the weblog into the students' assessment:

> This is why I eventually decided to grade the weblogs. Thanks to the blogs I got to know my students from a different perspective and realized the great potential of some of the quieter students (excerpt from the teacher's project evaluation).

This example underlines the importance of creating different learning environments in the geography classroom in order to create opportunities for every student to perform to their best ability.

9.10 Benefits for Educators

The results of this study show that reflective thinking differs among students of the same class, among lessons of the unit and between different classes. Several factors were found to account for these differences in reflective thinking, such as reference to everyday experiences, the implementation of key concepts, student gender and learning environment.

Summing up these findings, the aim was to develop a stage model of reflective thinking in the context of global education (Brendel, in press). This reflective model enables educators to easily determine the level of reflective thinking in students' written work. In addition, the model offers strategies and methods for targeted promotion of each student's individual reflective thinking performance.

With this model at hand, educators are equipped with a guideline for differentiated instruction regarding reflective thinking, which ideally should lead to the opportunity to provide individual support for each student according to their reflective thinking ability.

A profound consideration of the students' reflection on geographical issues entails a new focus on content in geography lessons. In the context of Education for Sustainable Development (ESD), *competences to think in systems* and *competences of assessment* are crucial for students to handle this geographical knowledge appropriately (Appelt and Siege 2008). As reflective thinking is an essential part of those competences, it can be regarded as a key competence in order to promote powerful geographical knowledge of our students and should gain more importance and attention in geography classrooms.

References

Andrusyszyn, M., & Davie, L. (1997) Facilitating reflection through interactive journal writing in an online graduate course: A qualitative study. *International Journal of E-Learning and Distance Education, 12*(1), 103–126. Retrieved from http://www.ijede.ca/index.php/jde/article/view/266/427

Appelt, D., & Siege, H. (Eds). (2008). Orientierungsrahmen für den Lernbereich Globale Entwicklung im Rahmen einer Bildung für nachhaltige Entwicklung: Ergebnis des gemeinsamen Projekts der Kultusministerkonferenz (KMK) und des Bundesministeriums für wirtschaftliche Zusammenarbeit und Entwicklung (BMZ). Berlin: Bundesministerium für wirtschaftliche Zusammenarbeit und Entwicklung.

Bain, J., Ballantyne, R., & Packer, J. (1999). Teachers and teaching: Theory and practice using journal writing to enhance student teachers' reflectivity during field experience placements. *Teachers and Teaching: Theory and Practice, 5*(1), 51–73.

Brendel, N. (in press). Reflexives Denken im Geographieunterricht. *Eine empirische Studie zur Bestimmung von Schülerreflexion mithilfe von Weblogs im Kontext Globalen Lernens*. Münster: Waxmann.

Carrington, S., & Selva, G. (2010). Critical social theory and transformative learning: Evidence in pre-service teachers' service-learning reflection logs. *Higher Education Research and Development, 29*(1), 45–57.

Catling, S. (2015). Geography in England's primary pre-service teacher education: Challenges and possibilities. In *IGU-CGE/UCLIOE Geography Education Conference*. London: UCL Institute of Education.

Chappell, A. (2006). Using the 'grieving' process and learning journals to evaluate students' responses to problem-based learning in an undergraduate geography curriculum. *Journal of Geography in Higher Education, 30*(1), 15–31.

Chen, N., Wei, C., Wu, K., & Uden, L. (2009). Effects of high level prompts and peer assessment on online learners' reflection levels. *Computers & Education, 52*(2), 283–291.

Dabbagh, N., & Kitsantas, A. (2011). Personal learning environments, social media, and self-regulated learning: A natural formula for connecting formal and informal learning. *The Internet and Higher Education, 15*(1), 3–8.

Dewey, J. (1997). *How we think*. Mineola: Dover Publications.

El-Dib, M. A. B. (2007). Levels of reflection in action research. An overview and an assessment tool. *Teaching and Teacher Education, 23*(1), 24–35.

Fögele, J. (2015). Geographical thinking through key concepts? Results of a symbiotic in-service teacher training course. In *IGU-CGE/UCLIOE Geography Education Conference*. London: UCL Institute of Education.

Gebhardt, H., Glaser, R., Radtke, U., & Reuber, P. (Eds.). (2007). *Geographie: physische Geographie und Humangeographie*. Heidelberg: Elsevier Spektrum Akad. Verl.

Gibbs, G. (1988). *Learning by doing: A guide to teaching and learning methods*. London: FEU.

Haigh, M. (2001). Constructing Gaia: Using journals to foster reflective learning. *Journal of Geography in Higher Education, 25*(2), 167–189.

Harrison, M., Short, C., & Roberts, C. (2003). Reflecting on reflective learning: The case of geography, earth and environmental sciences. *Journal of Geography in Higher Education, 27* (2), 133–152.

Hatton, N., & Smith, D. (1995). Reflection in teacher education: Towards definition and implementation. *Teaching and Teacher Education, 11*(1), 33–49. Retrieved from http://www.sciencedirect.com/science/article/pii/0742051X9400012U

Healey, M., & Jenkins, A. (2007). Kolb's experiential learning theory and its application in geography in higher education. *Journal of Geography, 99*(5), 185–195.

Henderson, K., Napan, K., & Monteiro, S. (2004). Encouraging reflective learning: An online challenge. In Atkinson, R., McBeath, C., Jonas-Dwyer, D., & Phillips, R. (Eds.), *Beyond the comfort zone: Proceedings of the 21st ASCILITE Conference* (pp. 357–364), Perth.

Hramiak, A., Boulton, H., & Irwin, B. (2009). Trainee teachers' use of blogs as private reflections for professional development. *Learning, Media and Technology, 34*(3), 259–269.

Hsieh, S., Jang, Y., Hwang, G., & Chen, N. (2011). Effects of teaching and learning styles on students' reflection levels for ubiquitous learning. *Computers & Education, 57*(1), 1194–1201.

Huffaker, D. (2004). The educated blogger: Using weblogs to promote literacy in the classroom. *First Monday, 9*(6), 1–5.

Hung, C. (2014). The use of weblogs in citizenship education: A theoretical analysis of the integration of weblogs and the English citizenship curriculum. *E-Learning and Digital Media, 11*(4), 350–362.

Jackson, P. (2006). Thinking geographically. *Geography, 91*(3), 199–204.

Kember, D., Jones, A., Loke, A., McKay, J., Sinclair, K., Tse, H., & Yeung, E. (1999). Determining the level of reflective thinking from students' written journals using a coding scheme based on the work of Mezirow. *International Journal of Lifelong Education, 18*(1), 18–30. Retrieved from http://www.tandfonline.com/doi/abs/10.1080/026013799293928#.VmV2Tl6714k

KMK and BMZ (Eds). (2015). Orientierungsrahmen für den Lernbereich Globale Entwicklung: im Rahmen einer Bildung für nachhaltige Entwicklung. Ein Beitrag zum Weltaktionsprogramm "Bildung für nachhaltige Entwicklung" (2nd ed.). Retrieved from http://www.globaleslernen.de/sites/default/files/files/link-elements/or_abschlussentwurf_din_a_4_final_21.07.15.pdf

Kolb, D., & Fry, R. (1975). Towards an applied theory of experiential learning. In C. Cooper (Ed.), *Theories of group processes*. London: Wiley.

Konak, A., Clark, T., & Nasereddin, M. (2014). Using Kolb's experiential learning cycle to improve student learning in virtual computer laboratories. *Computers & Education, 72*, 11–22.

Lambert, D. (2015). Powerful Disciplinary Knowledge and the Future 3 Curriculum. Keynote-Speech presented at Deutscher Kongress für Geographie. Deutsche Gesellschaft für Geographie and Geographisches Institut der Humboldt-Universität zu Berlin, Humboldt Universität zu Berlin.

Mansvelder-Longayroux, D., Beijaard, D., & Verloop, N. (2007). The portfolio as a tool for stimulating reflection by student teachers. *Teaching and Teacher Education, 23*(1), 47–62.

Mayring, P. (2000). Qualitative content analysis. Forum qualitative Sozialforschung/Forum. *Qualitative Social Research, 1*(2).

Mitchell, D. (2015). Supporting geographical thinking in the classroom—the teacher and curriculum control. In *IGU-CGE/UCLIOE Geography Education Conference*. London: UCL Institute of Education.

Morrison, K. (1996). Developing reflective practice in higher degree students through a learning journal. *Studies in Higher Education, 21*(3), 317.

Naghdipour, B., & Emeagwali, O. L. (2013). Assessing the level of reflective thinking in ELT students. *Procedia—Social and Behavioral Sciences, 83*, 266–271.

O'Loughlin, E., & Wegimont, L. (2002). Maastricht global education declaration. European strategy framework for improving and increasing global education in Europe to the year 2015. Online. www.coe.int/t/dg4/nscentre/Resources/Publications/GE_Maastricht_Nov2002.Pdf. Accessed 24 October 2014.

Oner, D., & Adadan, E. (2011). Use of web-based portfolios as tools for reflection in preservice teacher education. *Journal of Teacher Education, 62*(5), 477–492.

Overby, A. (2009). The new conversation using weblogs for reflective practice in the studio art classroom. *Art Education, 62*(4), 18–24.

Ozdemir, N. (2015). An experimental study related to application of the Kolb's learning cycle in geography lectures. *Hacettepe University Journal of Education, 30*(2), 81–98.

Park, C. (2003). Engaging students in the learning process: The learning journal. *Journal of Geography in Higher Education, 27*(2), 183–199.

Paterson, C., & Chapman, J. (2013). Enhancing skills of critical reflection to evidence learning in professional practice. *Physical therapy in sport: Journal of the Association of Chartered Physiotherapists in Sports Medicine, 14*(3), 133–138.

Pauw, I., & Beneker, T. (2015). The future in geographical thinking. In *IGU-CGE/UCLIOE Geography Education Conference*. London: UCL Institute of Education.

Qualifications and Curriculum Authority (QCA) and Department for Children, Schools and Families. (2007). The national curriculum statutory requirements for key stages 3 and 4. Retrieved from http://www.teachfind.com/qcda/national-curriculum-statutory-requirements-key-stages-3-and-4-september-2008-pdf

Richardson, W. (2010). *Blogs, wikis, podcasts, and other powerful Web tools for classrooms* (3rd ed.). Thousand Oaks, California: Corwin.

Roberts, M. (2013). Learning through enquiry. In *Speech presented at the conference "Powerful Knowledge: To what extent is this idea applicable to school geography?"*. London: UCL Institute of Education. Retrieved from https://www.youtube.com/watch?v=DyGwbPmim7o

Schön, D. (1983). *The reflective practitioner: How professionals think in action*. New York: Basic Books. Retrieved from http://www.loc.gov/catdir/enhancements/fy0832/82070855-d.html

Seow, T. (2015). How geography as a school subject is defined in Singapore. In *IGU-CGE/UCLIOE Geography Education Conference*. London: UCL Institute of Education.

Sharma, P. (2008). Student experience of using weblogs: An exploratory study. *Learning, 12*(3).

Su, Y. (2014). An investigation into the contents and aspects of college students' reflective thoughts during field experience. *Innovations in Education and Teaching International, 52*(3), 322–334.

Taylor, L. (2009). *GTIP think piece: Concepts in geography*. Retrieved from http://geography.org.uk/gtip/thinkpieces/concepts

The Geographical Association's national curriculum review group. (2012). *Thinking geographically*. Retrieved from http://www.geography.org.uk/download/GA_GINCConsultation12ThinkingGeographically.pdf

Timmins, F., Murphy, M., Howe, R., & Dennehy, C. (2013). 'I Hate Gibb's Reflective Cycle 1998' (Facebook©2009): Registered nurses' experiences of supporting nursing students' reflective practice in the context of student's public commentary. *Procedia—Social and Behavioral Sciences, 93*, 1371–1375.

UNESCO. (2014). *Roadmap for implementing the global action programme on education for sustainable development*. Retrieved from UNESCO website: http://unesdoc.unesco.org/images/0023/002305/230514e.pdf

van Manen, M. (1977). Linking ways of knowing with ways of being practical. *Curriculum Inquiry, 6*(3), 205–228.

Weichhart, P. (2003). Physische Geographie und Humangeographie—eine schwierige Beziehung: Skeptische Anmerkungen zu einer Grundfrage der Geographie und zum Münchener Projekt einer "Integrativen Umweltwissenschaft". In Heinritz G (ed) Münchener Geographische Hefte (Vol. 85, pp. 17–34). Integrative Ansätze in der Geographie—Vorbild oder Trugbild? Münchner Symposium zur Zukunft der Geographie, Eine Dokumentation.

West, K. (2008). Weblogs and literary response: Socially situated identities and hybrid social languages in english class blogs. *Journal of Adolescent and Adult Literacy, 51*(7), 588–598. Retrieved from http://www.jstor.org/stable/40012384

Wopereis, I., Sloep, P., & Poortman, S. (2010). Weblogs as instruments for reflection on action in teacher education. *Interactive Learning Environments, 18*(3), 245–261.

Yang, S. (2009). Using blogs to enhance critical reflection and community of practice. *Educational Technology and Society, 12*(2), 11–21.

Young, M. (2008). *Bringing knowledge back in: From social constructivism to social realism in the sociology of education*. London: Routledge.

Young, M. (2011). What are schools for? *Educação, Sociedade and Cultura, 32*, 145–155.

Young, M. (2013). Bringing knowledge back. In *'Powerful knowledge': To what extent is this idea applicable to school geography?* London: UCL Institute of Education. Retrieved from https://www.youtube.com/watch?v=r_S5Denaj-k

Zeichner, K., & Liston, D. (1985). Varieties of discourse in supervisory conferences. *Teaching and Teacher Education, 1*(2), 155–174.

Chapter 10
Geographical and Spatial Thinking in the Swedish Curriculum

David Örbring

10.1 Introduction

In 2011 a new curriculum was introduced in Sweden for its state schools—including primary, lower secondary and upper secondary schools (Skolverket 2011a). This revised curriculum made reference to subject specific abilities. The geography curriculum for state schools includes four subject specific abilities for students, which are designed to express the subject as taught to students from year 1 to year 9. In the research reported in this chapter, these subject specific abilities have been investigated and critically assessed. Firstly, the curriculum intentions of policy makers and curriculum developers are explored, with a particular focus on the view of knowledge that sits behind these stated subject specific abilities; secondly, these subject specific abilities are set in relation to thinking geographically and spatial thinking.

Findings about curriculum intentions, and the views of knowledge that lie behind the stated subject specific abilities which express these intentions, are noteworthy—specifically in relation to geographical and spatial thinking. This chapter investigates the views of knowledge advanced in the curriculum for secondary schools in Sweden with a specific focus on geographical learning and its curriculum representations. I begin by discussing the view of knowledge that permeates the state school curriculum and its particular structure, and continue by presenting the subject specific abilities for geography, concluding by considering their effects on the nature of geographical and spatial thinking expected of learners in Swedish schools.

D. Örbring (✉)
Department of Educational Sciences, Lund University, Lund, Sweden
e-mail: david.orbring@uvet.lu.se

C. Brooks et al. (eds.), *The Power of Geographical Thinking*,
International Perspectives on Geographical Education,
DOI 10.1007/978-3-319-49986-4_10

10.2 Views of Knowledge in the Swedish Curriculum

The 1994 state school curriculum in Sweden was based on *Skola för bildning* (Läroplanskommittén 1992), which presented significant changes in the ways in which knowledge was viewed in Swedish schools. Significantly, the curriculum had progressed from being process-oriented to being goal- and results- oriented (Englund et al. 2012). The government-sponsored study *Skola för bildning* (Läroplanskommittén 1992) described how this revised view of knowledge in the curriculum was to be implemented, outlining the new responsibilities for teachers. In 2011 Swedish schools received a new curriculum and syllabuses (Skolverket 2011a); the change of curricula that occurred in this year could be described as a structural reform rather than a shift of approach concerning the views of knowledge (Örbring forthcoming). Therefore, the ways in which knowledge is conceptualised in *Skola för bildning* (Läroplanskommittén 1992) are also the bases for the reform of the curriculum in 2011.

The term *bildning* is relevant in relation to the curriculum reforms of 2011—it also helps us to answer questions such as 'what is the purpose of education?' and 'what role does education play in the development of young learners?' Knowledge, as described in *bildning*, is recognised as having been the foundation of the Swedish school curriculum for almost 200 years, but with significant reinforcement in the 1994 curriculum (Läroplanskommittén 1992). Historically, there are different concepts of educational aims and practices connected to *bildning*—such as classical education, general education, and scientific and technical education. These have influenced the concept of *bildning* in Swedish education today; specifically the term refers to developing a variety of *forms* of knowledge that are not superficial or narrowly interpretive of the content of subject disciplines—but which also form part of the development of students' character, values and personality. In this way schools are expected to educate students to become more independent and reflective. Part of the concept of knowledge in the school is familiarity, and this has a clear link to how *bildning* affects the view of knowledge in schools. In the process of modern education, the following aspects are also of interest:

- Teachers and students are expected to achieve a historical understanding of how knowledge has developed.
- The school will offer a variety of educational paths and measures of success.
- Teachers should prevent socially constituted hierarchies of knowledge and understand that narrowly conceptualised fields of knowledge may become a barrier to knowledge growth.

So, what is the definition of knowledge currently applied in Swedish schools? The concepts of knowledge, as expressed through the Swedish school curriculum, were expanded during the 1990s to include more than just cognitive elements and to embrace social and practical aspects of knowledge forms. The conceptualisation of knowledge shifted from *inlärning* to *lärande* (Läroplanskommittén 1992)—specifically, knowledge could no longer be interpreted as something on the 'outside' to be 'transferred' to students, but instead it was to be seen as something that occurs in relation to human/student experience and actions. Essentially, the gaining of knowledge in Swedish schools (as presented to the student) is about the interplay between what the learner wants to achieve, the knowledge already possessed and the problems faced. Knowledge in Swedish schools is also constructive; it is not about a physical depiction of the world, but of a world socially created by humans. Importantly for geographers it is conceptualised as a way for us to make the world intelligible. Knowledge is also considered to be contextual and functional —that is, a tool for humans to understand and solve problems in different contexts (Läroplanskommittén 1992). A further expectation of the state school curriculum in Swedish schools is the development of *kunskapande* persons—this is an expression of the desire for education not only to represent a process of developing knowledge in young learners, but also to be about problem solving and drawing conclusions. *Kunskapande* is therefore also linked to the development of reflective thinking, communication and creativity (Läroplanskommittén 1992).

Four aspects of knowledge are considered central to its definition in Swedish schools—these are facts, understanding, skills and familiarity. These are believed to be interactive with each other and mutually dependent (Läroplanskommittén 1992). Fact is the informative aspect of knowledge, or the knowledge of information. Understanding is the ability to interpret and explain—including the ability to comprehend the meaning of something, or to recognize purposes. Additionally it is taken to include the realisation that the same phenomenon can be understood in different ways. Skills are about exercising knowledge, knowing what to do and how you can accomplish something. Familiarity is connected to review, linked to tacit knowledge and sensory experiences. For example, when we "feel" that something is going on. This form of knowledge usually becomes noticeable in evaluations.

10.3 Structure of the Swedish Curriculum

The curriculum for state schools in Sweden is divided into three parts (Swedish National Agency for Education 2011):

(1) Fundamental values and tasks
(2) Overall goals and guidelines
(3) Syllabus and knowledge requirements.

The Swedish National Agency for Education has been tasked to describe this structure and to articulate how the parts of the school curriculum are interconnected (Skolverket 2011b). 'Fundamental values and tasks' (part 1) are about the overall mission of the school system, as well as the knowledge and fundamental values that schools should help to develop in their students. 'Overall goals and guidelines' (part 2) are based on fundamental values and tasks, and are also divided into several sub-headings: norms and values, student responsibilities and influence, the school and the outside world, and assessment and grading. 'The purpose of overall goals and guidelines' is to guide the teachers and schools about how teaching should be carried out. This part also includes goals connected to knowledge—these goals are the basis for the syllabuses in the different subjects. The last part (3) is 'syllabus and knowledge requirements', which includes syllabuses in all subjects in state schools. The syllabus in each subject is divided into 'purpose', 'core content' and 'knowledge requirements'. The abilities, as with the aims of subject, are spelled out in the purpose of the syllabuses and are therefore also subject specific. The core content is more specific about what to teach. However, it does not specify for the teacher how much of the different contents should be taught, and teachers may add content freely, as long as they follow parts 1, 2 and the purpose of the subject. The knowledge requirements are what teachers must use for assessment and grading.

In upper secondary school, the curriculum structure is described in *Läroplan, examensmål och gymnasiegemensamma ämnen för gymnasieskola 2011* (Skolverket 2011d). As in most countries, a series of Education Acts (skollagen) form the basis for setting the statutory conditions for all education practice. Secondary regulations (gymnasieförordningen) are passed to make the Education Acts more 'concrete' in terms of how schools interpret and action them. The curriculum then describes values, mission, goals and guidelines for schools. Each programme also has goals of examination (examensmål), which will form the basis for the planning of education. Lastly, the syllabuses describe for each subject a subject plan. These subject plans include different courses that the students can attend and are seen as a comprehensive expression of curriculum intentions. The same structure is also valid for the curriculum for state schools (Swedish National Agency for Education 2011) without the goals of examinations. Education Acts take precedence over all other documents, while secondary regulation, the curriculum and goals of examination concretize the Acts in practice. Syllabuses regulate teaching, but also the goals of examinations, with the curriculum governing the teachers' planning of teaching. Student subject specific abilities in upper secondary school are specified in the syllabuses.

10.4 The Representation of Student Abilities in the Swedish Curriculum

The Swedish National Agency for Education is leading the process of developing the curriculum. The Ministry of Education has assigned them this task, and provides a structural framework that the National Agency for Education must follow. However, beyond this legal framework, the Swedish National Agency for Education is expected to use educational research and feedback from teachers in the process of writing the curriculum (Örbring, forthcoming). In drafting the curriculum of state schools they must present a framework where all subject syllabuses include statements on the development of students' abilities. All syllabus writers must therefore determine a number of subject specific abilities that should be written into the curriculum. Each subject is then entered into the overall curriculum framework. A new concept encapsulated in the reform of the education policy documents in 2011 has been the expression of the concept of subject specific abilities. This was conceived to capture, or at least describe, the views of knowledge that exist throughout all the school phases of primary and secondary education (Örbring, forthcoming). However, the theoretical background of knowledge in *Skola för bildning* (Läroplanskommittén 1992) is also the basis for this reform.

A series of interviews (Örbring, forthcoming) undertaken during the research that contributes to this chapter, show that there are influences to the process of drafting the curriculum—including consideration of the role of competences (for example, see DeSeCo Symposium 2003) and of different taxonomies of education (for example, see Anderson and Krathwohl 2001). The Swedish National Agency for Education (Skolverket 2010) has previously described the relationship between competence and knowledge. They refer to knowledge in the Swedish curriculum in very similar ways to the key competences stated by the EU and OECD. They also refer to a definition of competences as 'ett kunnigt och engagerat deltagande och handlande i en viss praktik' (Skolverket 2010: 15), which can be translated as 'a knowing engagement and participation in certain [educational] practices'. In comparison to taxonomies, the view of knowledge shown in the four 'aspects of knowledge' is not hierarchical, which means that one cannot put facts before understanding, skills and familiarity—these aspects should instead be seen as interdependent (Carlgren 2009: 15).

The term 'ability' is complex, and the nuances of its meaning in Swedish can easily become lost in translation. According to a Swedish encyclopaedia of education (Egidius 2006: 132), the term 'ability' means: To be able to cope with a specific task, or tasks. However, ability has taken on a new, revised definition in the reform of school documents in 2011 (Örbring, forthcoming), which attempts to capture the view of knowledge in *Skola för bildning* (Läroplanskommittén 1992). One definition of ability provided by the Swedish National Agency for Education is presented in the text *Kunskapsbedömning i skolan: praxis, begrepp, problem och möjligheter*; which translates from the Swedish as 'forms of knowledge covered by the broad knowledge concept of ability'. In this publication the concept of ability is used for different forms of knowledge. Synonymously with the concept of ability, the word knowledge is also used for *all* forms of knowledge (Skolverket 2011c: 6). In short, this means that the term ability covers knowledges (plural) and is used synonymously with the singular term 'knowledge'.

'Ability' is sometimes defined with reference to 'potential for performance'. It is unclear, however, how such ability might be accurately measured, with respect to performance. The term is neither synonymous with 'achievement' nor with 'attainment', and refers rather to what a student is capable of, rather than 'what they have proved themselves able to do' (Wallace 2008). A noteworthy aspect is also to look at the definition of the concept of ability in a dictionary related to research in geography education: '(It) describes the capacity to perform given tasks or skills…' (Butt 2000: 1). However, such definitions become even more complicated if you take Barrow's (1999) assertions about 'Higher nonsense' into consideration. He believes that much of the consideration of ability in education represents a careless use of the concepts of both ability and skill, where skill is often used to describe abilities. In conclusion, the Swedish National Agency of Education (2011: 32) state that the term 'ability' is to be used to express the concept of knowledge in the curriculum in a broad sense, asserting that it is inappropriate to be considered in combination with 'skills'. The reason for this, according to the Agency, is that abilities are synonymous with knowledge and that skills are just one of four parts of knowledge—the other parts being facts, understanding and familiarity.

10.5 Abilities in Geography

The descriptors applied to 'abilities in geography' from the Swedish curriculum are as follows:

> Teaching in geography should essentially give students the opportunities to develop their ability to:
>
> - Analyse how natural processes and human activities form and change living environments in different parts of the world,
> - explore and analyse the interaction between people, society and nature in different parts of the world,
> - make geographical analyses of the surrounding world, and evaluate the results by using maps and other geographical sources, theories, methods and techniques, and
> - assess solutions to different environmental and development issues based on considerations concerning ethics and sustainable development (Swedish National Agency for Education 2011: 150–151).

Defining specific abilities in geography, as is the case for most subjects, is difficult. To begin with, it is important to define and isolate which abilities are 'geographical', and what separates and differentiates geography from other subjects when considering the nature of (geographical) knowledge. In doing so it is arguably important to acknowledge other conceptual frameworks that relate to geographical abilities, such as thinking geographically and spatial thinking. An English translation of the Swedish curriculum in geography (Swedish National Agency for Education 2011) might consider the term *förmågor* to be best represented as *abilities*. It is intriguing to consider what lies behind this particular translation, and to reflect on how *förmågor* may be related to abilities in an international context (Örbring, forthcoming). In the Swedish context, one of the guiding concerns for curriculum makers has been to isolate the nature of subject specific abilities—indeed, the focus has been to deliberate on which subject specific abilities every citizen needs to learn (Swedish National Agency for Education 2009). These considerations are noteworthy in the light of recent research into GeoCapabilities (Lambert et al. 2015) which has, in part, sought to investigate what students need to learn as powerful knowledge in geography.

10.6 Thinking Geographically

A concept that has been debated in different ways over the years by geography educationists, but with slightly different foci, is the nature of *thinking geographically.* In the manifesto '*A different view*' (Geographical Association 2009) the concept of thinking geographically is emphasized; it is again pursued by Morgan in *Debates in Geographical Education* (Lambert and Jones 2013). Morgan traces the debate about geographical thinking back to Sir Halford Mackinder's founding ideas about modern geography: 'Not the collection of useless information about places but a trained capacity for thought [...] to be able to picture the world's dramatic spectacle on a stage' and 'To think geographically is to have a trained capacity to construct a mental map to see patterns, recognise relationships, to see movement...' (Lambert and Jones 2013: 275). Another aspect of *thinking geographically* may be linked to what Matthews and Herbert (2008) refer to as *an integrated approach.* This means that geography may constitute a bridge between nature and society, or indeed the arts and sciences, and may also be divided into three parts—'space, place and environment'. Morgan argues that Matthews and Herbert's definition of geography is related to these three parts: "The 'essence' of geography is the shaded area where the three concepts overlap as 'an integration of spatial variation over the Earth's surface with the distinctiveness of places and interaction between people and their environments'" (Lambert and Jones 2013: 277).

To understand what thinking geographically means, I have also considered Jackson's (2006) definition of geographically thinking. Jackson attempts to determine what geographical thinking means by describing four key concepts. These four key concepts are:

- Space and place
- Scale and connection
- Proximity and distance
- Relational thinking.

There has been much written about space and place [see, for example, Massey (2005) and Tuan (1974)], and geographical thinking describes, to some degree, the relationship between these terms. Different spatial scales, such as national and international, are often used in everyday geographic thinking. However, Jackson contributes ideas about various ways of looking at scale in the context of 'zooming in' and 'zooming out'. One example is that decisions taken at the local level can have global impact, and that global processes can have an impact at the local level. The third concept is about proximity and distance. Jackson highlights his belief that one's experience of closeness and distance is important, in the sense that events that occur far away can feel 'close by' and that new technology can make the sense of distant events appear more immediate. The fourth concept of relational thinking, according to Jackson, concerns how people may think of similarities and differences—this kind of thinking creates structures about 'us' and 'them', or indeed East and West.

Within the context of Jackson's definitions of geographical thinking sits a question about what separates geography, or geographical thinking, from that of other subject disciplines. Susan Hanson's concept of geographical advantage can be applied to help us strive for a definition of what geography is. Her concepts include consideration of the:

- Relationships between people and the environment;
- importance of spatial variability (the place-dependence of processes);
- processes operating at multiple and interlocking geographic scales; and
- integration of spatial and temporal analysis (Hanson 2004: 720).

10.7 Thinking Geographically in the Swedish Curriculum

Abilities in geography, which also summarize the view of knowledge in geography in the Swedish curriculum, may be connected with geographical thinking. This is particularly the case in consideration of:

- Integrated geography
- Interactions between humans, society and nature.

Phrases such as *how natural and human activities form and change* and *explore and analyse the interaction* are examples of integrated geography. In these phrases the intentions are to interpret both human and physical geography together. The interaction between people and environment is also visualised here. Part of this involves the application of geographical methods and theories as tools—such as maps, geographic information systems (GIS) and field studies. *Thinking geographically* is part of this ability.

The further stated ability in the syllabus is to 'assess solutions to different environmental and development issues based on considerations concerning ethics and sustainable development' (Swedish National Agency for Education 2011: 150–151)—although this could be seen as falling outside the description of geographical thinking. Instead, one may use geographical thinking to develop this ability (or, put differently, geographical thinking may be necessary to use this ability). I would go so far as to say that powerful knowledge in geographical thinking is the basis for working with this ability.

Comparing Jackson's key concepts with the stated abilities in the Swedish geography syllabus (see Table 10.1) provides some noteworthy points of similarity and difference. None of the words used by Jackson to describe 'thinking geographically' are present in the Swedish conception of geographical abilities. However, one might also argue that *all* of Jackson's key concepts are present in the abilities—they are just described differently, and translate in a more literal sense, the range of abilities in geography. It is perhaps a question of what is considered implicit and explicit.

Another way of analysing this is to compare the abilities with geographical advantage (see Table 10.2).

Table 10.1 Jackson's (2006) key concepts, as applied to the Swedish geography syllabus

	Space and place	Scale and connection	Proximity and distance	Relational thinking
Analyse how natural processes and human activities form and change living environments in different parts of the world	Implicit	Implicit	Implicit	Implicit
Explore and analyse the interaction between people, society and nature in different parts of the world	Implicit	Implicit	Implicit	Implicit
Make geographical analyses of the surrounding world, and evaluate the results by using maps and other geographical sources, theories, methods and techniques, and	Implicit	Implicit	Implicit	Implicit
Assess solutions to different environmental and development issues based on considerations concerning ethics and sustainable development	Implicit	Implicit	Implicit	Implicit

Table 10.2 Concept of geographical advantage, as applied to the Swedish geography syllabus

	Relationships between people and the environment	The importance of spatial variability (the place-dependence of processes)	Processes operating at multiple and interlocking geographic scales	The integration of spatial and temporal analysis
Analyse how natural processes and human activities form and change living environments in different parts of the world	Explicit	Implicit	Implicit	Implicit
Explore and analyse the interaction between people, society and nature in different parts of the world	Explicit	Implicit	Implicit	Implicit
Make geographical analyses of the surrounding world, and evaluate the results by using maps and other geographical sources, theories, methods and techniques, and	Explicit	Implicit	Implicit	Implicit

(continued)

Table 10.2 (continued)

	Relationships between people and the environment	The importance of spatial variability (the place-dependence of processes)	Processes operating at multiple and interlocking geographic scales	The integration of spatial and temporal analysis
Assess solutions to different environmental and development issues based on considerations concerning ethics and sustainable development	Explicit	Implicit	Implicit	Implicit

In this comparison one can see that some concepts are more explicit than others—particularly the relationship between people and the environment. The other parts of Hanson's (2004) definition are more implicit, that is to say these concepts could be included if teachers interpret the abilities with this in mind.

10.8 Spatial Thinking

The National Research Council report *Learning to Think Spatially* (2006) describes spatiality as being important both in the study of science and in everyday life. It claims that by using spatial thinking people may more easily define, understand and seek solutions to problems. Here spatial thinking is particularly linked to cognition.

It could be argued that spatial thinking is at the core of geographical thinking. Peet (1998) argues that, 'the synthetic core of geography is a study of nature-society interrelations' (Peet 1998: 2) and he describes spatiality as something that permeates geography—as witnessed by the spatial expression of interactions between nature and society. Graves (1982) also highlights spatiality as a core geographical concept, dividing it into three overlapping aspects central to the study of geography. These three aspects are:

- Spatial location
- Spatial distribution
- Spatial relations.

In *Learning to think spatially* (National Research Council 2006), the concept of spatial ability is described as follows: 'Spatial ability is conceptualized as a trait that a person has and as a way of characterizing a person's ability to perform mentally such operations as rotation, perspective change, and so forth' (National Research Council 2006: 26). The authors also refer to three categories of spatial ability: 'spatial perception, mental rotation, and spatial visualization'. Spatial ability is conceived of as a broader concept—spatial thinking, which consists of: 'space, representation, and reasoning'.

Baker et al. (2015: 120) have recently sought to define the nature of spatial and geospatial thinking. In doing so they proclaim that spatial thinking is: 'A set of abilities to visualize and interpret location, position, distance, relationship, movement, and change through space. *Spatial thinking and reasoning* involve cognitive processing of spatial data. This locational, positional, and measurement data is encoded and stored in memory, and can be represented externally by visualizations.' Additionally they define geospatial thinking as: 'A specialized form of spatial thinking that is bound by Earth, landscape, and environmental scales. *Geospatial reasoning skills* are higher-order cognitive processes that provide a means to manipulate, interpret, and explain information, solve problems or make decisions at geographic scale.' Such definitions are noteworthy in comparison to those of Ishikawa (2013) who attempts to distinguish spatial ability from geospatial ability, but believes that both concepts have a common point of reference in 'spatiality'. He argues that the term 'geospatial' links what is spatial with the specific substance of geography. Ishikawa therefore strives to consider how spatial ability relates to geospatial thinking—for instance, he shows that spatial ability is a part of geospatial thinking, but also that some aspects of geospatial thinking do not have a clear connection with spatial ability. The importance of distinguishing between knowledge and reasoning related to geospatial thinking is apparent. For some tasks, students need both knowledge and reasoning in geospatial thinking, while other tasks may be managed with only spatial reasoning (Ishikawa 2013). Additionally a study was recently published claiming that spatial ability is important and unique when it comes to children's ability to develop creativity (Kell et al. 2013).

10.9 Spatial Thinking in the Swedish Curriculum

Spatial and geospatial thinking are both identifiable in the Swedish curriculum. The specific abilities in the Swedish curriculum in state schools include certain terms that are noteworthy to compare:

- **Analyse** how natural processes and human activities **form and change living environments** in different parts of the world,
- **explore and analyse** the **interaction** between people, society and nature in different **parts of the world**,
- make **geographical analyses of the surrounding world**, and **evaluate the results by using maps** and other **geographical sources, theories, methods and techniques**, and
- assess solutions to different environmental and development issues based on considerations concerning ethics and sustainable development.

The words in bold text provide examples of when spatial thinking and geospatial thinking could be considered to be an important part of teaching and learning the stated geographical abilities. The ability 'to visualize and interpret location, position, distance, relationship, movement, and change trough space' could be seen in the words in bold text, and also in some parts 'by Earth, landscape, and environmental scales.' Spatial thinking and geospatial thinking are implicit as tools when working with these abilities. Therefore, it depends on the teachers' interpretation of the abilities whether or not spatial and geospatial thinking becomes part of the teaching and learning process.

10.10 Conclusions

The views of knowledge in Swedish schools contain a wide range of aspects, such as the four aspects of knowledge and *bildning*. These views of knowledge are supposed to be captured in the subject specific abilities and also cover tacit, theoretical and practical knowledge. The term ability is broad and is used to cover aspects of knowledge in a particular subject. Abilities are subject specific in the Swedish curriculum.

The statements of educational abilities—and subsequently the related view of knowledge of geography in Swedish state schools—include *thinking geographically* and *geographical advantage*, both of which are expressed implicitly in curriculum documents. How teachers interpret these stated abilities is therefore significant, as this will determine how thinking geographically is represented in the teaching process. Spatial thinking forms an implicit part of the geography syllabus—it is not highlighted, but is something that depends on how teachers interpret the syllabus.

Some noteworthy questions follow from this conclusion—is it important for spatial thinking to be made more explicit? Or should it form part of the teachers' professional judgement as to the ways in which it is represented in the (formal) curriculum, and then taught in schools? Geography teachers should be able to interpret the implicit parts of the formal geography curriculum with reference to geographical and spatial thinking in their teaching. However, previous studies have shown that geography teaching can still be traditional in many state schools in Sweden (Molin and Grubbström 2013):

> selective traditions in geographical education are strong, resulting in a focus on country-related knowledge and map-reading skills. Both teachers and students seem unclear about what other subject-specific skills geography teaching provides. Furthermore, students have difficulty achieving a high level of geographic reasoning. The authors argue that a subject-specific language in geography is important in both teaching and assessment. They stress that students need more practice in geographic reasoning (Molin and Grubbström 2013: 142).

A further issue, of greater significance, is that one third of geography teachers in state schools in Sweden have received no formal education in geography (for example, teachers in the social sciences in secondary schools often do not identify themselves as geography teachers). Geography is regularly seen as a 'handmaiden' subject which helps other subjects with tools and materials, rather than being seen as a discrete subject in its own right (Bladh 2014: 160–162). Bladh (2014: 167) comes to the conclusion that it is a considerable challenge for Swedish schools to promote geography as a subject, and in so doing create a further integration of the relationship between nature and society. Perhaps a more explicit expression of geographical thinking and/or spatial thinking could help teachers with these challenges?

References

Anderson, L., & Krathwohl, D. (2001). *A taxonomy for learning, teaching, and assessing: a revision of Bloom's taxonomy of educational objectives*. New York: Longman.

Baker, T., Battersby, S., Bednarz, S., Bodzin, A., Kolvoord, B., Moore, S., et al. (2015). A research agenda for geospatial technologies and learning. *Journal of Geography, 114*(3), 118.

Barrow, R. (1999). The higher nonsense: Some persistent errors in educational thinking. *Journal of Curriculum Studies, 31*, 131–142.

Bladh, G. (2014). Geografilärare och geografiundervisning i den svenska grundskolan. Några delresultat av en enkätstudie, Geografiska Notiser, 4.

Butt, G. (2000). *Continuum guide to geography education*. London: Continuum.

Carlgren, I. (2009). Kunskapssynen I 90-talets läroplanskonstruktion. In I. Carlgren, E. Forsberg & V. Lindberg (Eds.), *Perspektiv på den svenska skolans kunskapsdiskussion*. Stockholm: Stockholms universitets förlag.

DeSeCo Symposium. (2003). Contributions to the second DeSeCo Symposium. Neuchâtel, Swiss Federal Statistical Office (SFSO), Geneva, Switzerland, February, 11–13 2002.

Egidius, H. (2006). *Termlexikon i pedagogik, skola och utbildning*. Lund: Studentlitteratur.

Englund, T., Forsberg, E., & Sundberg, D. (Eds.). (2012). *Vad räknas som kunskap?: läroplansteoretiska utsikter och inblickar i lärarutbildning och skola*. Stockholm: Liber.

Geographical Association. (2009). *A different view: A manifesto from the geographical association*. Sheffield: Geographical Association.

Graves, N. (1982). *New UNESCO source book for geography teaching*. Longman: Harlow.

Hanson, S. (2004). Who are 'we'? An important question for geography's future. *Annals of the Association of American Geographers*, 715.

Ishikawa, T. (2013). Geospatial thinking and spatial ability: An empirical examination of knowledge and reasoning in geographical science. *Professional Geographer, 65*, 636–646.

Jackson, P. (2006). Thinking geographically. *Geography, 91*(3), 199–204.

Kell, H., Lubinski, D., Benbow, C., & Steiger, J. (2013). Creativity and technical innovation: Spatial ability's unique role. *Psychological Science, 24*, 1831–1836.

Lambert, D., & Jones, M. (Eds.). (2013). *Debates in geography education*. Abingdon: Routledge.

Lambert, D., Solem, M., & Tani, S. (2015). Achieving human potential through geography education: A capabilities approach to curriculum making in schools. *Annals of the Association of American Geographers, 105*(4), 723.

Läroplanskommittén. (1992). *Skola för bildning: huvudbetänkande*. Stockholm: Allmänna förl.

Massey, D. (2005). *For space*. London: Sage.

Matthews, J., & Herbert, D. (2008). *Geography: A very short introduction*. Oxford: Oxford University Press.

Molin, L., & Grubbström, A. (2013). Are teachers and students ready for the new middle school geography syllabus in Sweden? Traditions in geography teaching, current teacher practices, and student achievement. *Norwegian Journal of Geography, 67*, 142–147.

National Research Council. (2006). *Learning to think spatially: GIS as a support system in the K-12 curriculum.* Washington DC: National Academies.

Örbring, D. (forthcoming). Dissertation at Lund University. Department of Educational Sciences.

Peet, R. (1998). *Modern Geographical Thought.* Oxford: Blackwell Publishers.

Skolverket. (2010). *Redovisning av uppdrag att utarbeta nya kursplaner och kunskapskrav för grundskolan och motsvarande skolformer m.m.* Skolverkets rapport nr U2009/312/S. Dnr 2008:741. Stockholm: Skolverket

Skolverket. (2011a). *Läroplan för grundskolan, förskoleklassen och fritidshemmet.* Stockholm: Skolverket. Available online: http://www.skolverket.se/publikationer?id=2575

Skolverket. (2011b). *Planering och genomförande av undervisningen: för grundskolan, grundsärskolan, specialskolan och sameskolan.* Stockholm: Skolverket.

Skolverket. (2011c). *Kunskapsbedömning i skolan: praxis, begrepp, problem och möjligheter.* Stockholm: Skolverket.

Skolverket. (2011d). *Läroplan, examensmål och gymnasiegemensamma ämnen för gymnasieskola.* Stockholm: Skolverket.

Swedish National Agency for Education. (2011). *Curriculum for the state school system, the pre-school class and the recreation centre (Lgr11).* Stockholm: Swedish National Agency for Education (Skolverket). Available online: http://www.skolverket.se/publikationer?id=2687

Tuan, Y. F. (1974). *Topophilia: A study of environmental perception, attitudes, and values.* Englewood Cliffs, NJ: Prentice-Hall.

Wallace, S. (2008). *A dictionary of education.* Oxford: Oxford University Press.

Chapter 11
GIS and the Power of Geographical Thinking

Mary Fargher

11.1 Introduction

For almost a quarter of a century, technical, cost and training challenges permitting, GIS in geography has been used to carry out decision-making and problem-solving on the 'whys of where.' The type of geography underpinning GIS has always had spatial thinking at its heart. In particular, pedagogy associated with GIS in schools has focused on using a constructivist enquiry-based model to support geography. Whilst the latter is well-suited to a conventional GIS as an information system it is true to say that the thinking underpinning it can limit geographical thinking to a focus on the spatial and certain types of decision-making, problem-solving, site analysis and prediction. Whilst teaching and learning through these can be very successful on some levels, it is important for teachers and educators to consider which types of geographical thinking are more suited to and which are more difficult to achieve through GIS. At the same time, the world of geospatial technologies is changing. GIS has moved online, digital geographical information is now available on a wide range of platforms including networks, laptops, tablets and phones.

Research evidence reflecting the pedagogical benefits of using GIS to enhance school geography is well established (Bednarz and Bednarz 2004; Doering and Valetsianos 2008; Favier and van der Schee 2012). Where GIS is being used successfully it is promoting spatial literacy; supplementing fieldwork; and enhancing pupils' visualisation of geographical phenomena in increasingly interactive digital environments often through geographical enquiry. Whilst some GIS remains difficult to access in terms of cost, technical access and teacher training, the proliferation of free online GIS on a range of platforms (PC, laptop, network, mobile) has started to make digital geographical information more accessible. More

M. Fargher (✉)
UCL, Institute of Education, London, UK
e-mail: m.fargher@ucl.ac.uk

C. Brooks et al. (eds.), *The Power of Geographical Thinking*,
International Perspectives on Geographical Education,
DOI 10.1007/978-3-319-49986-4_11

teachers and pupils are beginning to gain access to GIS in the school curriculum (Fargher 2013). At a time when such developments in GIS are accelerating, it is important that the development of geographical thinking through GIS should reflect the full educational potential that geographical information systems now have to offer but also that geography teachers and educators are fully familiar with the types of advantages and limitations on geographical thinking that GIS can also bring. In the next section the discussion begins with close consideration of the ways in which geographical knowledge can be constructed in GIS and the types of geographical thinking associated with it.

11.2 Constructing Geographical Knowledge in GIS

It is true to say that spatial thinking not geographical thinking per se, lays at the core of GIS knowledge construction. In their report: 'Learning to think Spatially' (2006) the US National Research Council (NRC) presents the argument that space as represented in GIS can provide a conceptual and analytical framework within which geographical data can be represented and interpreted. The type of spatial thinking advocated by NRC is underpinned by understanding the world through very specific digital spatial structures rather than thinking geographically in its broader sense. Spatial thinking can be perceived in terms of how an individual visualises their surroundings, how they orientate themselves and navigate within that space and how they make informed judgements about relationships between elements of their environment.

Spatial visualisation is the cumulative effect of the processes by which an individual perceives their immediate and more distant surroundings. Spatial orientation focuses on an individual's abilities to place them within space and to navigate around it. Spatial relations include the ability to recognize the significance of places and to make informed judgements about connections between them. Each of these elements of spatial thinking can be internalized by an individual as components of their unique spatial cognition. Hart and Moore (1973) argue that:

> Spatial cognition is the knowledge and internal cognitive representation of the structure, entities, and relations of space; in other words, the internalized reflection and reconstruction of space in thought (Hart and Moore 1973, p. 248).

What is significant to the central discussion in this chapter is how thinking through GIS influences how an individual adapts their spatial cognition.

One of the most significant elements of using GIS in constructing geographical knowledge is through spatial query. At the most basic level, spatial querying could involve asking a simple set of questions of GIS data. Table 11.1 illustrates some simple examples (after Rhind 1992).

Rhind's classification of spatial querying in GIS shows both the kinds of questions that can be asked of spatial data in GIS and the nature of knowledge constructed as a result. His classification is significant for geography teachers wishing to use GIS to support the development of geographical thinking for a number of reasons. First, geographical data handling in a GIS is always closely bound to location which Milson et al. (2012) describe as the 'whys-of-where'. This refers to the ways in which GIS lends itself to 'What is at?' and Where is?' type questioning and the collection of geographical information. Second, collecting sets of data at different time periods can lead to useful geographical comparison of sites over time. Third, mapping locational data sets can enable teachers to construct geographical knowledge which leads to high quality spatial analysis. Fourth, using different sets of locational data can lead to useful comparison of different scenarios and modelling different alternatives for located geographical data sets. Table 11.2 illustrates the steps in a typical GIS-based enquiry.

Table 11.3 classifies rather more precisely, exactly how the construction and use of geographical knowledge may be explored with GIS according to one particular classification devised by Mark (1993). Mark classifies the type of knowledge produced through GIS geographically. He identifies three inter-related categories: *declarative*, *procedural* and *configurational* geographical knowledge. According to Mark's classification, declarative geographical knowledge is factual knowledge *about* geographical space. This may include facts about for example, names of locations and landforms. Table 11.3 can be used to classify the type of geographical knowledge that is constructed in a lesson supported by GIS; it can also be used as a

Table 11.1 Spatial querying in GIS (after Rhind 1992)

Question	Type of knowledge construction
What is at?	Inventory
Where is?	Monitoring
What has changed since?	Inventory and monitoring
What spatial pattern exists?	Spatial analysis
What if?	Modelling

Table 11.2 Steps in enquiry-based learning with GIS (after Malone et al. 2005)

Steps in enquiry	What to do	Type of knowledge construction
Ask geographical questions	Ask questions about your surroundings	Description
Acquire geographical resources	Identify data and information that you need to answer your questions	Monitoring
Explore geographical data	Turn the data into maps, tables, graphs and look for patterns and relationships	Inventory and monitoring
Analyse geographical information	Test a hypothesis, carry out map, statistical, written analysis	Spatial Analysis
Act upon your geographical knowledge	Reach geographical conclusions, inform a decision, solve a problem	Modelling

Table 11.3 Classification of GIS geography knowledge (after Mark 1993)

Declarative knowledge	Procedural knowledge	Configurational knowledge
Place (names), Location, spaces, distance, scale, vocabularies about place (e.g. region, disaster zone, settlement infrastructure)	→GIS navigational skills→ • Orientate • Pan • Zoom • Link • Co-ordinate • GPS	Recognising and interpreting Geographical patterns Mapping geographical patterns • Display • Measure • Attribute • Tabulate • Overlay • Input • Edit • Symbolize • Aggregate • Classify Analysis • Query • Interpolate • Statistical

basis for thinking about the opportunities and limitations of thinking geographically through GIS.

Declarative geographical knowledge may form the basis of the knowledge that individuals acquire, but, in reality, procedural and configurational transformation of that knowledge lead to deeper geographical thinking. Mark suggests that procedural geographical knowledge includes information which allows us to think about and navigate *through* geographical space. Using the basic elements of declarative geographical knowledge we think and act on information assimilated about landmarks and routes as we find our way from place-to-place. Mark defines configurational geographical knowledge as knowledge that allows us to formulate map-like information. Using declarative knowledge as a base and, perhaps, direct procedural geographical knowledge, we configure pattern, direction, orientation and hierarchy. For example, using the London Underground Map illustrates how we think about and mentally represent this type of configurational knowledge. The colour-coded topology of the map allows us to focus only on essential information: routes, line directions and 'anchor points' such as station interchanges. As a result, geographical knowledge is transformed into a new simplistically relational form.

It can be argued that an emphasis on relational geographical knowledge involves the more sophisticated level of geographical thinking and reasoning that have already been suggested. These include many potentialities: understanding geographical difference; relations between places; broader interpretation of human-environment interaction at a variety of scales; deepening spatial comprehension; and enhancing geography-rich spatial decision-making to name but a few. This section has made clear that as well as providing opportunities for spatial querying and geographical enquiry, constructing knowledge through GIS places a number of significant restrictions on thinking geographically in the broader sense. The next section considers the origin of this critique further by examining the origins and development of the field of critical GIS.

11.3 Critical GIS: A Review

Despite the fact that the first incarnation of GIS in the field of GIScience was developed in the form of the Canada Geographic Information System (CGIS) as early as the mid-1960s, disquiet regarding the philosophical ramifications of GIS only began to be widely debated by geographers in the early 1990s. Some critical (mainly human) geographers, who until that point had seemed to be quite content about the development of this new technology began to air caution about the proliferation of GIS technologies and the implications of this. Their concerns were varied, but in particular, they feared what they saw as a retrograde step back to the quantitative/scientific ways of thinking geographically that was dominant during the emergence of spatial science in the 1960s and 1970s. Not only this, they were beginning to become more aware of the social implications of mass-produced geographical information (Schuurman 2000).

The main critique of GIS has always been that conventional GIS cannot adequately represent the world through its narrow parameters. In particular, the view of the world portrayed as an independent Cartesian grid system through conventional GIS on which social processes of geography are located. Proponents argued that human geography in particular was far too complex and varied a discipline to be framed and understood through the positivist science of conventional GIS. This was strong criticism, particular in relation to the idea that concepts of space (both geometric and relative) and the spatial reasoning most often resulting from using GIS can narrow worldviews. The renewed stirring of unease in the 1990s was only the start of a protracted spat between advocates of GIS and those rather cautious of its roots and implications. By the mid-1990s, the criticisms which began to surface and develop were not just based on suspicion or lack of technical knowledge about computing. Mainly, concerns were voiced on the basis of two over-arching criticisms: the positivist origins of GIS; and the possibilities of questionable ethics behind its intentions (Schuurman 2000; Sui 1994).

The focus of critics of GIS on more philosophical concerns and implications for geographical thinking was to prove the most notable at that time. Dissatisfaction with GIS settled for academics on its potential to limit ways in which individual users could think geographically (Lake 1993; Sui 1994). For some, the 'language of GIS' with its emphasis on spatial and attribute data seemed to shape a kind of sheer, aesthetically-pleasing representation where potentially 'fuzzy social knowledge' was rejected in favour of the smooth clean lines of vectors and polygons. In summary, then, the dangers levelled by critics were seen to stem from the inherently positivist nature of GIS which was blamed for producing a technology which could only quantify and not qualify. It could be argued that unquantifiable social phenomena could not be accommodated in a geographical information system. As Schuurman (2000) identifies, critics' opinions were encapsulated by a strong view that GIS was: 'by implication, a means of limiting the proliferation of epistemologies in geography' (Schuurman 2000, p. 580).

The polemic which continued to build during the early to mid-1990s was debated more fully in 1995 with the publication of *Ground Truth: The Social Implications of Geographic Information Systems* edited by John Pickles. This collection of essays challenged the technology, the ways of thinking underpinning it and the many social implications of using it to construct knowledge. Whilst there can be no argument about the coincidence of the 'birth of GIS' with the 'spatial turn' in geographical thinking in the 1960s, the belated criticism which emerged about GIS nearly three decades later was based not only around arguments about its quantitative pedigree but were also based on negative claims about its positivist origins. Coincidentally, by the mid- 1990s, often, separate to the critique debate on GIS, other human geographers were already embarked on developing the very opposite of a limited and narrowed approach to geographical thinking. The rising status of GIS in some academic geography departments coincided with the culmination of a decade of re-thinking academic geography. If GIS really did have the power to threaten the proliferation of theoretical approaches developing within the subject, this was the most damning criticism of all.

A more constructive trend within geography followed 'Ground Truth' with the development of the field of critical GIS. The latter's more progressive research agenda partially opened up a forum where advocates and sceptics debated rather more productively over the relative merits and drawbacks of GIS. Pivotal to the inception of critical GIS was the role of the US National Center for Geographic Information Analysis (NCGIA) (Schuurman 2006). NCGIA began to direct discussion towards a fuller analysis of the social ramifications of how GIS represents people, space and environments (O'Sullivan 2006). The epistemological debates that raged and developed as a response or indeed a reaction to the earlier quantitative revolution had matured.

Marxist, feminist and humanistic geographers had already prepared and justified their theoretical approaches to practising and delivering their own brands of geographical knowledge (Pickles 2006). Other responses to these early critiques of GIS included Kwan's work on feminist GIS and emotional geographies (Kwan 2002, 2007); Brown and Knopp on queer GIS (2008); alternatives to western-centric GIS (Crampton and Krygier 2005) and works on 'democratising the technology' (Elwood 2002; Sieber 2007). Kwan offers a related and poignant argument for developing a wider critique of geospatial technologies (Kwan 2007). Her discussion forms a renegotiation of the 'meanings of GIS at the intersection of science, art, and subjectivities'. In particular, Kwan offers an imaginative analysis of the ways in which GIS can be used as a 'medium of self-expression and a means of resistance'

(Kwan 2007, p. 25). Kwan develops this idea in relation to contemporary world issues such as 'natural' disasters, globalization and international conflict in ways that are relevant to using GIS to teach about such 'issues' in geography in education. There is also an interesting parallel here with the ethics of using GIS—through connecting young people with the moral implications of using GIS, and developing their understanding of the 'subjectivities' of the knowledge produced through the social and political practices of GIS (Kwan 2007).

Leszcynski (2009) provides a useful summary of recent developments in critical GIS. Her argument goes some way to moving forward the debate between GIScience and critical GIS because she emphasises that modern GIS has changed in some respects from its early version that emerged from the tenets of the quantitative revolution (Crampton 2009). She examines the 'discursive separation' between GIScience and critical human geography, emphasising the lack of common language between the two schools of thought that often leads to a misunderstanding of each other's aims and practices. Her discussion emphasises the important role that the poststructuralist critique of GIS has played in contributing to the more ethical mapping that has emerged through critical GIS of late. Leszcynski contextualises this approach clearly offering a persuasive argument for mapping through GIS that provides a counter-view to the 'imperial cartographic conventions' of Cartesian perspectivalism. She implies an inherent responsibility for all mapping whether digital or otherwise to include opportunities for representing difference in geography. In: *Theorising with GIS*, Pavlovskaya (2006) takes this critique of GIS further, moving on from the binary opposition of quantitative GIS versus qualitative human geography to a broader critique.

11.4 Geographical Thinking Through New Geospatial Technologies

In connection with new developments in GIS, Elwood (2009) discusses the major changes that have occurred in terms of new geospatial technologies, particularly earth viewers such as Google Earth. This has included a major change in the way that some geospatial data can be accessed. As Elwood discusses:

> In the world of geospatial technologies, change is afoot. In the past five years, we have seen the emergence of a wide array of new technologies that enable an ever-expanding range of individuals and social groups to create and disseminate maps and spatial data (Elwood 2009, p. 256).

As Elwood points out, these online applications share some commonality with traditional GIS in terms of digital storage; retrieval and visualisation, there are actually few other similarities.

Elwood (2009) addresses the plethora of complexity that new 'spatial technologies' offer. She argues that GIS are more than just the sum of their technological capacity. Instead, she contends that they are:

> Digital systems for storing and representing spatial information; they are complex arrays of social and political practices; *and* they are ways of knowing and making knowledge (Elwood 2009, p. 257).

The increased availability and the easier technical use of technology such as Google Earth raise important questions about how the information within is created. Dodge and Perkins (2009) contend that:

> Different theoretical approaches may be deployed to interrogate the significance of these powerfully affective visual technologies, and indeed it can be argued that the democratisation of satellite image accessibility is itself part of a significant epistemological shift (Dodge and Perkins 2009, p. 498).

In their consideration of the political implications of how Google Earth and satellite imagery was used in New Orleans after Hurricane Katrina, Crutcher and Zook (2009), examined the 'mapped cyberscape' and offered an analysis of the role of imagery in contributing to geographical thinking and the use of Google Earth place marks in a series of mash-ups posted online. Their analysis offers a fascinating insight into the nature of the racial composition of an area and the 'post-Katrina cyberscape in Google Earth.' In a similar way, the 'Digiplace' that Zook and Graham (2007) describe is invisible for those without the economic means of access or the cognitive and technical capacities to access it. Zook and Graham (2007) argue:

> DigiPlace represents the situatedness of discrete individuals straddling virtual and physical realities, rather than any sort of shared, objective, and fixed reality (Zook and Graham 2007, p. 8).

In her examination of the use of Google Earth in school education: *Digging into Google Earth: An analysis of 'Crisis in Darfur*, Parks (2009) argues that treating earth viewers such as Google Earth as a field in which questions about the world can be initiated and answered is one particular way of using GIS effectively. Parks's argument also focuses on this idea of interaction between the use of digital technologies and the political and social processes at work in the world. She contends:

> Armed with such information, world citizens might be more apt to pressure their governments to formulate proactive as opposed to reactive foreign policies, to understand world conflicts as more than the primitivism or pathology of tribal warfare, and to help push the historical forces and power hierarchies that shape the planet into bold relief (Parks 2009, p. 544).

Crutcher and Zook (2009) also argue that the use of satellite imagery with online user comments can (as quoted in Dodge and Perkins 2009): 'create highly differentiated connections between places and cyberspace.' In a similar argument, Parks (2009) discusses the benefits and disadvantages of representation of 'real events' through GIS. In her examination of 'Crisis in Darfur' she describes the illusion (my interpretation) of the idea that perhaps the closer 'zoomed view' in Google Earth brings with it a deeper understanding. But does it? She also considers what the 'god's eye view' through satellite imagery may also encourage in terms of interpretation and understanding. She argues:

> The structural interplays between the far and near that undergird Google Earth's Crisis in Darfur project are helpful in that they represent the potential to refigure key terms of humanitarianism. The iconographies of suffering are not reduced merely to images of people, but are placed in dynamic alternations with satellite images and graphics that emphasize the territorial and the geopolitical (Parks 2009, p. 544).

In a similar way, but in a discussion about digital maps in: 'Plotting the personal: Global Positioning Satellites and interactive media' (2001), Parks also contests the meanings of place knowledge created through geospatial technologies. For example, in her consideration of GPS tracking technology, she argues that there is a fusion between the personal experience and the social landscapes 'portrayed' and accessed through GIS. In a later examination: Digging into Google Earth: An analysis of 'Crisis in Darfur' (2009), Parks explores the implications of what she views as stereotyped views of Africa presented through Google Earth and the 'disaster capitalism' (term coined by Naomi Klein). She contends: 'In short, Google Earth is not 'a view from nowhere'—it is the view from a company with enormous visual capital' (Parks 2009, p. 542).

Kingsbury and Jones (2009) support the idea that GIS introduces digital worlds that are also only 'knowable' in a perspectival way i.e. they are very much open to individual interpretation despite some superficial evidence to the contrary, as discussed below. In a rather impassioned view, they argue:

> A virtual globe composed of surveyed panoramas, sober rationalization, dystopic control, and transparent order – or, even, as a tool for participation and empowerment – we undersell its capacities as an alluring digital peep-box, an uncertain orb spangled with vertiginous paranoia, frenzied navigation, jubilatory dissolution, and intoxicating giddiness (Kingsbury and Jones 2009, p. 502).

Whilst Kingsbury and Jones's analysis is perhaps a more colourful critique of digital portals such as Google Earth, it does emphasis a need for the further development of a discourse about the new spaces of cultural politics that is continuing to open up through this new type of web-based GIS.

Kingsbury and Jones (2009) analyse a series of exchanges about the tools used through Google Earth to 'observe' the crisis in Darfur. They argue:

> Google Earth's profane illuminations, wherein objects gaze back at us as we disappear (see Doel, 2006) or get lost in meanings that are at once excessive and insufficient, demands attending to new geometries of power in scopic regimes (Kingsbury and Jones 2009, p. 508).

There is an intoxicating element to this new found ability to interact with and think about large scale geographic space through GIS. Kingsbury and Jones describe this (2009) as 'surfing under the influence of Google Earth.' They make a number of claims about Google Earth's role as a 'digital peep-box', its *failure* to 'reveal a visible and legible world' and its non-Foucauldian role as panopticon (Kingsbury and Jones 2009, p. 505). In discussing Lammermen and Bergsma's 'peep box concept' (2006), Kingsbury and Jones (2009) contend that:

> The peep-box boundary equates to Google Earth's 3D geo-referenced scene, the peep-box faces stand in for Google Earth's digital pictures and bit-maps, the cellophane represents

Google Earth's atmospheric lights' and 'clear blue skies', while the peephole mimics the interface between users and the 3D scene' (Kingsbury and Jones 2009, p. 505).

There is also the question of the partisan nature of the ways in which Google censors some material uploaded into Earth and Maps and not other material, despite the often open appearance of debate about this on the Internet. Zook and Graham (2007) argue that Google Earth's DigiPlace is some considerable distance down the road in becoming the de facto digital globe. They contend that there: 'remains much to be done in understanding how new lived spaces are influenced by the intersections between culture, code and place' (Zook and Graham 2007, p. 18).

In connection with the hegemonies of GIS and maps as 'partisan place,' there is also the issue of the way in which GIS technology is being used as a tool of representation. The nature of the tools of construction within digital geographies becomes explicit when looking at the types of mosaicked pixels shown in satellite imagery. Dodge and Perkins make some important observations about the ramifications of these when they contend that:

Perhaps the biggest 'lie' of this seamless imagery is that it is constructed out of tiles from different times, which are mosaicked together to create a wholly artificial view devoid of cloud cover. Also as much imagery is captured in periods when skies are clear, and when vegetation growth is maximised and visually prominent, the result often obscures the built environment, and this diminishes the presence of people in the landscape (Dodge and Perkins 2009, p. 498).

In a discussion that pre-dates modern use of the virtual globes, Gelernter (1992) considers making worlds in digital space and the role of the individual as participant and 'geographical thinker' in a new visual environment. He emphasises a new kind of perspective on geographical thinking through GIS, one which is partly subjective and partly simulacra. He states:

A mirror world is some huge institutions moving, true-to-life mirror image trapped inside a computer-where you can see and grasp it whole. The thick, dense sub-world that encompasses you is *also* now an object in your hands. A brand new equilibrium is born (Gelernter 1992, p. 3).

Here are opportunities for considering geographical thinking as a *subjective* and *dynamic process* in ways that were not as feasible through traditional GIS but which have arguably become more accessible since the advent of new geospatial technologies. Kingsbury and Jones (2009) support the idea that GIS introduces digital worlds that are also only 'knowable' in a different perspectival way i.e. they are very much open to individual interpretation despite some superficial evidence to the contrary as discussed below. If 'DigiPlace' is not permanent or fixed space but is shaped by people places and processes in a way that is not only changing space within GIS but also material space outside then this has ramifications for knowing place, per se. Dodge and Perkins also argue that:

The visual virility and earthly representational richness of these images *can* also make recognition and interpretation of space much harder. The viewer may actually understand less of the structure of the place without the classification and clarity offered up by the professional eye of the cartographer (Dodge and Perkins 2009, p. 498).

This is one of the 'challenges of space' (Massey 2005) with GIS and why in geography and geography education, the way in which geographical knowledge is constructed still matters. It is important to note here that the metadata relating to GIS albeit map or satellite imagery metadata about data capture and resolution, for example, is usually hidden way below the surface of something like an earth viewer. To the untrained end user, this metadata is not usually accessible; for teachers using GIS this is ethically and educationally significant. Parks (2009) calls for caution regarding the potentialities of earth viewers such as Google Earth. She argues that:

> While Google Earth presents exciting new possibilities for integrating and accessing documents and audio-visual materials and providing them in geo-referenced fashion, information interventions such as Crisis in Darfur need to be discussed and evaluated carefully (Parks 2009, p. 544).

This section has considered the ways in which new cultural spaces opening up on the web raise other questions about geographical thinking through geographical information systems and the broader spectrum of geospatial technologies now available. The next section considers these implications further by examining the nature of geographical thinking through public participatory GIS (PPGIS).

11.5 Geographical Thinking Through Public Participatory GIS

One result of the many critiques of GIS has been a more socially-aware form of GIS which celebrates a broader range of geographical knowledges than the narrowly scientific ones most often associated with traditional GIS (Dunn 2007). The most championed movement to emerge within critical GIS has been public participatory GIS (PPGIS). Public participatory mapping usually involves using easier to use earth viewers or mapping applications such as Google Earth. For example, the Amazon Conservation Team (ACT) has used public participatory GIS to bring together indigenous geographical knowledge and GIS mapping technology (Tulloch 2007).

Though still not short of critical review by some sceptics, public participatory GIS is considered by some to be 'GIS for the people.' It embraces an approach where local issues drive the use of technology and where there is an emphasis on community involvement with GIS. Some of the projects that adopt a PPGIS approach have included urban regeneration and sustainable development e.g. the Worldfish research projects which were set up in the Aceh province of Indonesia to tackle environmental regeneration after the 2004 South Asia Tsunami (Worldfish 2013). In particular, PPGIS such as Worldfish differ from more traditional GIS in that they include more complex geographical information, often including indigenous geographical information. This in itself is an interesting aspect of PPGIS because it attaches importance to deep local knowledge.

Dunn describes the inclusion of indigenous geographical information in the following way:

> This is not technical knowledge but rather 'deep knowledge' which places cultural values on land and place, which is manifested in fuzzy, emotional and holistic terms (McCall and Minang 2005) and which may not fit neatly into the spatially precise demands of a GIS (Dunn 2007, p. 623).

In this way, PPGIS has a great deal to offer for teachers wishing to develop geographical thinking through geospatial technologies in the future through the digital mapping of more complex and less easily spatially-definable geographical knowledge.

11.6 Conclusions

This chapter has addressed the question: *What are the ramifications for geographical thinking of teaching and learning through GIS in both its more traditional guise and 'new GIS'*? In particular it has identified the advantages of using GIS to promote spatial thinking through visualisation, orientation and through spatial relations with regards to constructing declarative, procedural and configurational geographical knowledge. These processes have been identified as being beneficial in geographical thinking through GIS which focuses on locational knowledge, developing navigational skills and most significantly bringing together geographical knowledge and thinking in a deeper and more relational form.

The discussion has also highlighted the significance of this kind of spatial querying in supporting the constructivist enquiry-based model of teaching and learning in geography, this being to date still the most used strategy in GIS-based teaching and learning in school geography.

The chapter has also provided a comprehensive review of the critical GIS movement which has, since the 1990s clearly identified and worked towards the use of GIS which moves away from its positivist roots and towards GIS as a practice and culture which lends itself to more open-ended geographical thinking.

The discussion has also identified the deep significance of new geospatial technologies in developing geographical thinking particularly with regards to online applications and virtual globes which are beginning to shift GIS more towards the use of volunteered geographical information (VGI) and the opportunities that this may or may not provide for using more heterogeneous data in GIS particularly through public participatory GIS (PPGIS).

References

Bednarz, R. S., & Bednarz, S. W. (2004). Geography education: The glass is half full and it's getting fuller. *The Professional Geographer, 56*(1), 22–27.
Brown, M., & Knopp, L. (2008). Queering the map: The productive tensions of colliding epistemologies. *Annals of the Association of American Geographers, 98*(1), 40–58.
Crampton, J. W. (2009). Cartography: Maps 2.0. *Progress in Human Geography, 33*(1), 91.
Crampton, J. W., & Krygier, J. (2005). 'An introduction to critical cartography.' *ACME: An International E-journal for Critical Geographies, 4*(1), 11–33.
Crutcher, M., & Zook, M. (2009). Placemarks and waterlines: Racialized cyberscapes in post-Katrina Google Earth. *Geoforum, 40*(4), 523–534.
Dodge, M., & Perkins, C. (2009). The 'view from nowhere'? Spatial politics and cultural significance of high-resolution satellite imagery. *Geoforum, 40*(4), 497–501.
Doering, A., & Valetsianos, G. (2008). An investigation of the use of real-time, authentic geospatial data in the K–12 classroom. *Journal of Geography, 106*(6), 217–225.
Dunn, C. E. (2007). Participatory GIS—a people's GIS? *Progress in Human Geography, 31*(5), 616–637.
Elwood, S. (2002). GIS use in community planning: A multi-dimensional analysis of empowerment. *Environment and Planning A, 34*(5), 905–922.
Elwood, S. (2009). Geographic information science: New geovisualization technologies—emerging questions and linkages with GiScience research. *Progress in Human Geography, 33* (2), 256–263.
Fargher, M. (2013). Geographic Information (GI)—How could it be used? In D. Lambert & M. Jones (Eds.), *Debates in geography education*. London: Routledge.
Favier, T., & van der Schee., J. (2012). Exploring the characteristics of an optimal design for inquiry-based geography education with geographic information systems. *Computers & Education, 58*, 666–677.
Freundschuh, S. M., & Egenhofer, M. J. (1997). Human conceptions of spaces: Implications for GIS. *Transactions of GIS, 2*(4), 361–375.
Gelernter, D. (1992). *Mirror Worlds, or, the day software puts the universe in a shoebox: How it will happen and what it will mean*. USA: Oxford University Press.
Golledge, R. G., & Stimson, R. J. (1997). *Spatial behaviour: A geographical perspective*. New York: Guilford.
Hart, R. A., & Moore, G. (1973). The development of spatial cognition: A review. In R. M. Downs & D. Stea (Eds.), *Image and Environment*. Chicago: Aldine.
Kingsbury, P., & Jones, J. P., III. (2009). Walter Benjamin's Dionysian adventures on Google Earth. *Geoforum, 40*(4), 502–513.
Kwan, M. P. (2002). Is GIS for women? Reflections on the critical discourse in the 1990s. *Gender, Place and Culture: A Journal of Feminist Geography, 9*(3), 271–279.
Kwan, M. P. (2007). Affecting geospatial technologies: Toward a feminist politics of emotion. *The Professional Geographer, 59*(1), 22–34.
Lake, R. W. (1993). Planning and applied geography: Positivism, ethics, and geographic information systems. *Progress in Human Geography, 17*(3), 404–413.
Leszczynski, A. (2009). Quantitative limits to qualitative engagements: GIS, its critics, and the philosophical divide. *The Professional Geographer, 61*(3), 350–365.
Malone, L., Palmer, A. M., Voigt, C. L., Napoleon, E., Feaster, L. (2005). *Mapping our world : GIS lessons for educators*. Redlands: ESRI Press.
Mark, D. M. (1993). 'Human spatial cognition'. In D. Medyckyj-Scott & H. M. Hearnshaw (Eds.), *Human factors in geographic information systems*. London: Belhaven.
Massey, D. (2005). For space Sage. *Thousand Oaks CA*.
McCall, M. K., & Minang, P. A. (2005). Assessing participatory GIS for community-based natural resource management: Claiming community forests in Cameroon. *The Geographical Journal, 171*(4), 340–356.

Milson, A. J., Demirci, A., & Kerski, J. J. (Eds.). (2012). *International perspectives on teaching and learning with GIS in secondary schools*. New York: Springer.

National Research Council. (2006). *Learning to think spatially: GIS as a support system in the K–12 curriculum*. Washington, D.C: The National Academies Press.

O'Sullivan, D. (2006). Geographical information systems: Critical GIS. *Progress in Human Geography, 30*(6), 783–791.

Parks, L. (2009). Digging into Google Earth: An analysis of "Crisis in Darfur". *Geoforum, 40*(4), 535–545.

Pavlovskya, M. (2006). Theorising with GIS: A tool for critical geographies. *Environment and Planning A, 38*(11), 2003–2020.

Pickles, J. (1995). Representations in an electronic age: Geography, GIS, and democracy. In J. Pickles (Ed.), *Ground truth: The social implications of geographic information systems*. New York: Guilford Press.

Pickles, J. (2006). Ground Truth 1995–2005. *Transactions in GIS, 10*(5), 763–772.

Rhind, D. (1992). The information infrastructure of GIS. In *Proceedings of the 5th International Symposium on Spatial Data Handling* (Vol. 1, pp. 1–19). Columbia, SC, USA: Humanities and Social Sciences Computing Lab, University of South Carolina.

Schuurman, N. (2000). Trouble in the heartland: GIS and its critics in the 1990s. *Progress in Human Geography, 24*(4), 569–590.

Schuurman, N. (2006). Formalization matters: Critical GIS and ontology research. *Annals of the Association of American Geographers, 96*(4), 726–739.

Sieber, R. E. (2007). Spatial data access by the grassroots. *Cartography and Geographic Information Science, 34*(1), 47–62.

Sui, D. Z. (1994). GIS and urban studies: Positivism, post-positivism, and beyond. *Urban Geography, 15*(3), 258–278.

Tulloch, D. L. (2007). Many, many maps: Empowerment and online participatory mapping: http://www.firstmonday.org/ojs/index.php/fm/article/viewArticle/1620/1535. Last accessed 28th October, 2013.

Tversky, B., Bauer Morrison, J., Franklin, N., & Bryant, D. J. (1999). Three spaces of spatial cognition. *The Professional Geographer, 51*(4), 516–524.

Worldfish. (2013). Fisheries rehabilitation in post-tsunami Aceh: Issues & recommendations from a National Consultative Workshop. *NAGA, WorldFish Center Quarterly* (29), 3 & 4, Jul-Dec 2006. Available at: http://www.worldfishcenter.org/search/node/tsunami%20. Last accessed on 28th October, 2013.

Zook, M. A., & Graham, M. (2007). The creative reconstruction of the internet: Google and the privatization of cyberspace and digiplace. *Geoforum, 38*(6), 1322–1343.

Part III
Recontextualising Geographical Thinking

The third part of this book looks at the theme of geographical thinking within a broader context. The term geographical thinking is, in some ways, a hybrid term to describe something that happens when geographical knowledge and concepts are being used by an individual. The relationship between other concepts such as powerful knowledge and curriculum have been debated and described by the chapters herein. The term "geographical thinking" is profoundly educational, in that it is transformative and dynamic and offers a range of possibilities. The focus of the chapters in this book has been to offer a discussion about geographical thinking: what does it mean, how can we make it happen, what does it look like when it is taking place? The chapters in this part take a broader view of these questions, and reflect on some of the processes of recontextualisation that take place before geographical knowledge and concepts (and by extension geographical thinking) reach the classroom.

Any publication about education, pedagogy, curriculum and school subjects is to some extent concerned with recontextualisation. The presentation and transformation of knowledge to an appropriate medium for a classroom-based interaction requires that knowledge is recontextualised into an appropriate form. This book is no different. Part 1 has debated different ways of understanding the relationship between geographical thinking and powerful knowledge. Part 2 has taken a more pedagogical turn, focussing on what recontextualisation might look like in terms of teaching and learning in geography classrooms. In this part, the focus is on the recontextualisation that takes place outside of the classroom: therefore, the chapters in this part are concerned with the recontextualisation of knowledge by teachers, textbook authors and teachers within teacher education. Each chapter explores how geographical knowledges are processed and transformed before they are mediated and presented in the classroom. The process of mediation is important, as the knowledge that features in the classroom has already had to go through several processes in order to be presented to students. Our argument is that each of these processes has to be conscious of the aim of geographical thinking in order for this mediation to be effective.

This part consists of four chapters, written by Clare Brooks, Victor Salinas Silva, Andoni Arenas-Martija and Laura Ramirez-Lira, Felisbela Martins, and Simon Catling and Jongwon Lee. Examples are given from England, Chile and Portugal. As with other chapters, despite the national perspective offered the contribution of each is not limited to those countries—indeed the observations made about the processes of knowledge recontextualisation through teachers' expertise, alternative sources of knowledge, the teacher education process and the creation of textbooks are pertinent regardless of context. Indeed the focus on knowledge and thinking outside of the classroom enables us to see the contexts outlined in Part 2, and the theoretical ideas debated in Part 1 within some of the broader processes that affect and influence geography education.

Clare Brooks's chapter *International differences in thinking geographically and why 'the local' matters* starts the part by introducing this idea of recontextualisation within the context of geography teachers' understanding of geography, as an academic discipline, within England. Although her research was originally designed to understand different conceptions of geography, her findings highlight that social, historical and legacy issues are important mediators of geographical knowledge. Geographical knowledge, she argues, is subject to processes of recontextualisation before the active engagement of individual teachers in their particular school contexts. This sociological perspective on geography education is an important reminder of the "persistent grammar of schooling" and why educational systems are so resistant to change.

The theme of teachers' knowledge is continued by Victor Salinas Silva and his colleagues in their chapter entitled *Getting back to basics: Is the knowledge of school geography powerful in Chile?* This chapter not only explores the issues of what influences how geography is understood and taught in Chile, but also how the context of education is very different: focusing on rural schools and on how their location affects the disciplinary knowledge of teachers. One of the key findings that this chapter highlights is the distinction between where geographical knowledge comes from (and how it relates to everyday experience) and the ways in which teachers interact with that knowledge. Within these rural contexts, teachers draw upon their local environment, circumstances and experiences and use these effectively in the classroom to widen the conception of what geography means to young people. The process of geographical thinking here is rich!

Staying with the theme of teachers' geographical knowledge, Felisbela Martins looks at how teacher education can encourage new teachers to act as curriculum makers in her chapter: *Teaching to develop Geographical Thinking*. Martins's chapter is in two parts. The first part makes a theoretical argument, largely based on Portuguese education academics, as to the purposes of curriculum and how teachers should work with centrally prescribed curriculum contents. The argument is made strongly that teachers should not merely receive and deliver the geography curriculum, but should consider how it has to be adapted for the particular needs of individual classes. Martins's argument is that this is best achieved through a focus on what she describes as 'problem-situations', and in the second part of her chapter she offers an example of how this can be achieved for the topic of migration. The analysis of this example shows how the important themes of knowledge, and pedagogy, can come together to promote geographical thinking—exposure to this, in a pre-service context, can help teachers to develop their skills as curriculum developers.

The chapter *English geography textbook authors' perspectives on developing pupils' geographical knowledge and thinking* is the research of Simon Catling and Jongwon Lee. Based on a survey of geography textbook authors writing for English geography school textbooks, Catling and Lee explore the processes of recontextualisation that takes place in textbook construction. In particular they focus on the motivations and intentions as articulated by textbook authors. This chapter is a timely reminder that recontextualisation takes place at a number of levels and contexts and that the geography teacher is key to the selection, presentation and articulation of geography at the classroom scale. The influence of other teaching materials, of which textbooks are a key component, should not be ignored.

The focus on textbooks in the last chapter of this part links to one of the findings in the first chapter. Brooks reminds us that assessment, and in particular public examination specifications, are also hugely influential on how the curriculum is enacted in the classroom. The two chapters that are central to this part, both with their emphasis on teachers and teacher education, balance out these findings. Textbooks and examinations are important, but the quality of our geography teachers is key.

Chapter 12
International Differences in Thinking Geographically, and Why 'the Local' Matters

Clare Brooks

12.1 Introduction

In this chapter the theme of thinking geographically is explored in relation to teachers' subject knowledge and in particular their definitions of geography. This exploration is underpinned by an understanding that pedagogy is an expression of culture (Alexander 2001)—therefore how we teach about geography reflects how geography as a subject is valued and understood within that particular culture. Such an observation suggests that in order to develop and encourage geographical thinking, greater attention needs to be paid to how geography is defined locally.

There are two ideas that are relevant to this enquiry: the first is Bernstein's (1977) notion of recontextualisation, which helps us to understand what happens to knowledge when it is presented in the classroom. The second idea explores the relationship between school subjects and their academic parents, which can vary due to three key factors identified by Stengel (1997). Combined, these two ideas represent a model of the transformation of geographical knowledge within a pedagogical context: how teachers work with the subject to help their students to think geographically.

This chapter exemplifies this through data derived from pilot research conducted with geography teachers in England. The findings reveal that the relationship between the subject and how it is defined in the classroom context is not straightforward, but is influenced by the contexts that teachers work in—in particular how assessment regimes affect the curriculum, and the context and priorities of the individual departments. Such observations, whilst based on a small sample size, contribute further understanding to Stengel's model of how school subjects are related to their academic parents, and reveals other pertinent factors in the recontextualisation of knowledge.

C. Brooks (✉)
UCL, Institute of Education, London, UK
e-mail: c.brooks@ucl.ac.uk

C. Brooks et al. (eds.), *The Power of Geographical Thinking*,
International Perspectives on Geographical Education,
DOI 10.1007/978-3-319-49986-4_12

12.2 The Relationship Between Disciplines and School Subjects

School subjects have a variety of ways in which they can be related to their academic counterparts. In her analysis of this relationship, Stengel (1997) argues that the relationship between the two is contingent upon:

- the relative focus on academic, pedagogical, utilitarian and existential concerns;
- the extent to which the moral is allowed and encouraged; and
- the underlying view of knowledge.

These three factors are key because they relate to how disciplines are constituted within their own cultural setting, but also how the relationship between discipline and subject is contingent on how the latter is valued and understood within the particular education context. In other words, the approach recognises significant disciplinary and educational influences. School geography, therefore, may be valued differently in different places depending on local perceptions of the academic discipline and on how education is valued and constructed. For example, should geography be seen as a utilitarian academic subject of practical significance to how citizens interact with their local environment in one locale, then the school subject will be differently constructed in comparison to another locale where it is seen as more of a philosophical or descriptive discipline. This observation recognises that academic disciplines are not universally uniform in how they are viewed and valued. Geography as an academic discipline will have different emphases in different university departments, and, like its related school subject, is subject to "contestation and compromise" (Goodson 1987) at a variety of scales.

Stengel's categorisation also recognises different ideological approaches to education (see Rawling 2001). School subjects are socially constructed at a variety of levels: through the definition of national curricula, or public examinations, through school structures and organisations and also through the interpretation of the curriculum by individual teachers. Subject associations, other professional groups, textbook authors and resource developers go some way to defining the curriculum, and therefore the different priorities it reflects. However, it is at the scale of the classroom, where the subject has to come alive for students, that the expression of these interest groups can be seen. This research therefore focusses on the views of geography teachers who are at the central point of interpreting all these influences, and are subject to the combination of influences outlined above.

12.3 Subjects as Recontextualised Knowledge

The framework outlined by Stengel exemplifies how the process of preparing knowledge to be taught requires a variety of social constructions that take place by and beyond the individual teacher. Bernstein's idea of the pedagogic device and how this involves recontextualisation and reproduction is a useful way of understanding this process further (Bernstein 1977). Bernstein differentiated between everyday (or mundane) knowledge and esoteric knowledge: (or between the thinkable and the unthinkable: the not yet known), and identifies the pedagogic device as occupying the space between the two. The process of creating new knowledge is described by Maton (2014) as 'production'—distinctively different to the process of making knowledge available to others who do not have access to it.

Recontextualisation takes place at various levels of curriculum construction. Maton describes this as the "sites where knowledges from the field of production are selected, rearranged and transformed to become pedagogic discourse" (2014: 48). This field of recontextualising is created through official statements of a curriculum, such as through a National Curriculum, examination specifications or even textbooks. It is the rules of recontextualising that affect the transmission and acquisition of knowledge, and are made of an instructional discourse embedded within the regulative discourse (Bernstein 1990). In other words, by taking knowledge from the disciplines and making it available for students, it is subjected to transformation. Using the example of school physics, Bernstein explains:

> The rules of relation, selection, sequencing and pacing (the rate of expected acquisition of the sequencing rules) cannot themselves be derived from some logic internal to physics nor from the practices of those who produce physics. The rules of reproduction of physics are social, not logical, facts. The recontextualising rules regulate not only selection, sequence, pace and relations with other subjects, but also the theory of instruction from which the transmission rules are derived (1990: 185).

Bernstein's argues that these recontextualising rules are influenced by the classification and framing of the regulative discourse: that they are subjected to the status and value given to the subject, and subjected to its own grammar.

The notion of recontextualisation, as outlined by Bernstein, is useful because it highlights the curriculum work undertaken by teachers as part of a larger mechanism of transforming knowledge: "Changes in the theory of instruction can then have consequences for the ordering of the pedagogic discourse and for the ordering of pedagogic practice" (1990: 189). Bernstein also differentiates between the official recontextualising field (ORF) (that comes from official sources such as the State) and the pedagogic recontextualising field (PRF) (which may come from other sources that may influence pedagogy). What subsequently happens in the classroom can be understood by the evaluative rules of the pedagogic device.

It is therefore possible to see how knowledge from the discipline is transformed into classroom activities and discussion as part of the fields of recontextualisation and reproduction. The curriculum needs to be understood within the wider context of the ORF, and how that knowledge is originally created (distributive rules) and then used in the classroom (the evaluative rules). For example, as changes in education occur subjects have to realign themselves to the new context, redefining and reshaping their contribution to education: this will affect the recontextualisation of knowledge between the parent discipline and the school subject through the factors outlined by Stengel.

Taken together the recontexualisation of knowledge, and Stengel's relationship factors, are powerful ways of understanding how the curriculum gets enacted—and may help to explain how geography may be taught differently throughout the world. A focus on the recontextualisation of knowledge that focusses on the content of curriculum only may down play some of the other influential factors. A focus, for example, only on "powerful knowledge" may not take into account some of the other processes that occur in curriculum construction, and other localised influences such as the "pull" of particular assessment practices.

12.4 Exploring How Geography Is Defined

The research was inspired by an observation by Uhlenwinkel (2012) about the differences between the British use of the term "thinking geographically" and the American use of the term "spatial thinking". Uhlenwinkwel argues that such differences reveal conceptual as well as semantic variation. The aim of the research that followed was to gain a deeper understanding of how geography was defined as a subject and the significant influences on such definitions. The initial study was conducted with groups from four different countries (England, Chile, Singapore and Portugal), with a view to revealing different influences on how geography is understood in each locale. This chapter reports on the findings from the English group only, although Silvas et al's chapter in this section reports on the findings from the Chilean teachers.

The sample size of this pilot study is small, but the implications are believed to be significant. International research and dialogue is predicated on an assumption that there are some similarities in how geography is defined and valued across the globe. Should there be significant differences, then our understanding of research conducted in international contexts will also differ. There is already evidence that this is the case (see, for example, Butt and Lambert 2014). So, a greater appreciation of the nuances of these differences is vital for the effective transference between jurisdictions of geographical ideas and research findings. At the heart of this is a concern for what it means to 'think geographically', and whether the meaning of that phrase is shared internationally. The recent work on powerful knowledge (see, for example, Firth 2012; Roberts 2014; Young et al. 2014), which stems from the work of Michael Young, is predicated on an assumption that there can be some agreement (if contested) as to what that means for geography. As the chapters in Section One reveal, such an understanding is actually highly contextualised. To comprehend the power of geographical thinking it is important to understand the nuances of the influential factors in each of our national contexts.

The aim of this research was to understand what influences definitions of school geography but not to seek to define geography. The definitions that the research team were seeking to reveal were those held by geography teachers, as they are the key focus of the range of influences that can affect how geography is understood (both from the ORF and the PRF). It is recognised that individual teachers may have a personal definition of geography that is different to that expressed in the national context, but also that they will be aware of this and will be able to articulate these differences. It was hoped that findings from this study—combined with those from other contexts—will provide a robust basis for further international and large scale research in this area.

The pilot study was undertaken in Chile, England, Portugal and Singapore. Here the findings of the pilot study in one of the four countries that participated, England, is presented. Chapter 13 of this volume reports on the data collected in Chile. In each of the four countries, four to six experienced geography teachers were selected to participate. The participants needed to have at least three years' experience of teaching geography. The research team were also aware that some countries required teachers to teach more than one subject, and so felt it was important that all the participating teachers were specialist geographers, active within their subject communities.

The data collection consisted of two parts. The first part required participants to construct a hierarchical concept map, along the lines outlined by Seow (2013). The concept map was first modelled on a non-related example (in this instance we used President Barack Obama, as someone that most people would know something about), and then the participants constructed their own concept map on the topic of "geography". The topic was not defined any further (i.e., was not separated into school, academic or popular geography) but was left open to be interpreted by the participant. The concept maps were not included as part of the data analysis, but were used as a heuristic device to support the participants in the second part of the data collection.

The second part of the data collection was in the form of an individual interview. The interviews were set up at the participant's convenience and were audio recorded. The participants were encouraged to use their concept maps to help them answer three main questions. The questions were:

1. What is geography?
2. How is geography understood in this country?
3. How does school geography differ from what you have already said?

The questions were ordered in this way to enable the participants to discuss geography in a broader sense, if they wished to, and then to compare their personal perspective to how they perceived the national understanding of geography. The third question was designed to enable them to reflect and respond on whether their previous answers had been specific to school geography, and how school geography might differ to other representations they had discussed.

Interviewers were encouraged to ask follow up questions for clarification, but not to lead the participant. The results from the English geography teachers are discussed below.

12.5 Competing Discourses

In the discussions with each of the teachers who participated in the research different approaches and structures to the geography curriculum were revealed. Despite individual differences in layout and format (and specific content), the overall structure of a thematic approach was preferred by all the participants. Interview discussions revealed the influence of geographical concepts as a guiding principle for many of the participants. Such an outcome is not surprising. The English Geography National Curriculum, prior to 2008, was structured around geographical themes, and many school geography textbook series remain structured in the same way. Geographical concepts were introduced in the 2008 curriculum, and appear to have influenced how teachers articulate and think about learning in geography. The most recent geography national curriculum (introduced in 2014) has been marked by a return to knowledge, and this appears to have been expressed as a return to a thematic approach to curriculum construction. All the teachers discussed concepts such as space, place and scale (each of which were included in the 2008 national curriculum) as important in the aims of the curriculum—but the curriculum contents were expressed thematically.

The outline of geographical content was remarkably similar between each of the respondents, suggesting coherence in understanding and approach. The contents and discussion would also support the observation of the influence of the National Curriculum as a key part of the ORF, and also how the curriculum is interpreted within schools. However, an interesting difference emerged with a distinction between the discourses of this official version of the geography curriculum, referred to by the teachers as "traditional" and what they described as their own preferred approach.

For example, all the participants made a contrast between geography's ability to bring together the physical and human aspects of the subject—which they saw as important—with how it is expressed within official discourses, which they described as "traditional". The traditional approach appeared to be related to previous versions of the English National Curriculum which defined geography as a series of themes. The participants saw this version of geography as the dominant discourse and regarded it as a traditional approach, although they recognised that it lacked coherence. The description of this approach was contrasted to their own preferred approach which was more holistic and sought to combine both the physical and human aspects of the discipline, rather than to teach each in isolation.

For the participants in this research, school geography concepts as outlined in their concept maps, and as discussed in their interviews, were strongly related to the 2008 version of the geography national curriculum; the guiding concepts were the same as those articulated in that curriculum (for example, space and place). However, when these concepts were discussed in the interview no explicit reference was made to the National Curriculum, rather to a sense of what was seen as "good practice in geography education". This was not linked to any explicit reference to geography as an academic discipline, but was articulated as a recognition of the importance of the professional associations for geography teachers in influencing the curriculum (named explicitly as the materials from the Geographical Association and the Royal Geographical Society with IBG).

It is also interesting to note particular inclusions and omissions in relation to how school geography was defined by the English teachers. For example, the respondents used the term "space" (a key concept from the 2008 Geography National Curriculum) but did not mention "spatial". When talking about the specifics of the curriculum they referred to maps and sometimes to the importance of understanding scale, but there was no specific mention of geographical information systems or geospatial technologies. This might suggest that these terms and technologies have yet to enter into the day-to-day world of English geography teachers, or that they place more importance on low-tech approaches to learning about space. With such a small scale study it is not possible to determine the significance of such a finding, although it may be worthy of further investigation.

A similar subtle difference could be seen in relation to place. Teachers discussed the importance of learning about the local area and teaching about "place": they agreed that understanding "places" was important. However, they did not outline any rationale for which places should be studied (referring instead to resource availability as being a key decider in case study choice) and they were clear that they did not want a return to regional geography (an opinion expressed explicitly by two of the respondents).

In addition, all the respondents had a very clear idea of geography expressed at a macro level: emphasising in particular the coming together of understanding about the human/social and physical worlds. For example, one respondent described how geography was the only subject in the school curriculum that explored how social and physical science related to each other. This holistic perspective appeared to be shared by all the participants and was expressed as part of the key contribution of geography to the whole school curriculum. However, when asked how this vision was realised at a micro scale (for an individual scheme of work, or individual lesson), the participants were unable to see how this holistic vision of geography could be reflected here. It appeared that they understood the curriculum making process as being focussed on individual lessons which addressed small components, which were then built into themes, which the teachers "hoped" would build into a holistic picture of geography. The word "hoped" is emphasised here, as further discussion with the participants revealed that the extent to which this was effective, or was working, was unclear: they expressed it something they wanted to achieve but were not sure if this was happening. Indeed, as three of the participants pointed out, it is difficult to assess holistic understanding—something that is also not required in many public examination assessment schemas.

None of these observations are particularly surprising, reflecting as they do long standing issues that have concerned the English geography education community. However, the respondents also identified a further factor which had a significant influence on their curriculum choices: the context of the department, and its position of power and influence within the school.

12.6 The Influence of the School Context

One of the significant factors that was raised by each of the participants was the notion of internal subject competition within the school. In England, Geography is an optional subject at the age of 14. The introduction of the EBacc (English Baccalaureate) and Progress 8 (both national measures of school performance), has meant that students are encouraged to choose between History and Geography. Understandably this can generate a sense of competition between the two subjects: competition that is focussed on the recruitment of students, and which can influence how the department seeks to promote or market geography as an object of study. For example, one participant explained that geography needed to be seen as more relevant and academically rigorous than history for it to be more popular with the students. The teacher argued:

> History is more popular because it is seen as more rigorous, and so has a higher status amongst the brighter boys (T1).

However, this does not mean that rigour is equated with a subject being regarded as "better", as later in the interview the same teacher argued that geography needed to be seen as being more relevant to the students. When this response was questioned, in the light of History typically being regarded as an irrelevant subject, the teacher was unable to explain whether rigour or relevance was more important: what was important was being more attractive to potential students than history, which may mean different things for different students. This exchange revealed how school geography has to be understood in relation to other school subjects. The factors outlined in the exchange, rigour and relevance, can both be seen in the three features outlined by Stengel—which are key to how academic disciplines get recontextualised into school subjects. This finding suggests that these three factors can be determined by local factors such as internal competition. If the emphasis in an individual school is on relevance then school geography is likely to emphasise that in the local curriculum. At a school level, internal competition can mean redefining geography to place it in a more advantageous position: factors which may lead to focussing on particular elements of the discipline and its transformation in the school subject, due to the particular educational context.

This perspective is also reflected in the discussion around teacher responsibility in relation to defining the school geography curriculum. For example, one teacher said:

> Teachers shouldn't decide what is in the curriculum, that's not their job. Teachers should decide how to teach the curriculum (T2).

In response to this statement, the teacher was asked whose responsibility he thought it was, if not that of the teacher, to which he responded: "*Experts—like the RGS, or academics*" (T3).

This exchange highlights the teachers' lack of clear recognition of their responsibility towards the curriculum. Within England, the Geographical Association has sought to promote teachers as curriculum makers, emphasising the importance of their role in constructing a local curriculum. One aspect of being a curriculum maker is making appropriate selections from the academic discipline in the light of the particular group of students: so a curriculum in an urban school with a diverse population, might be different to that enacted in a rural school with a population from a similar class and ethnicity. This statement is also evocative of Smith and Girod's (2003) observation that curriculum authors are not always aware of individual students, and therefore published curricula require a degree of modification and interpretation for different student groups. However, discussions like the exchange above reflects Morgan's observation that teachers do not seem to feel that they are able to make these decisions anymore, deferring instead to an external undisclosed "expertise".

> Crudely stated, geography teachers may experience some confusion as to whether their job is to transmit geographical knowledge, prepare autonomous learners who are able to 'learn how to learn', or promote social cohesion through notions of global citizenship (Morgan 2011: 91).

This is reflected in the perceived conflict between the traditional and holistic curriculum as outlined above. This was reflected in every interview, where the traditional approaches (which were not always clearly defined) were placed in contrast to a holistic approach to teaching geography. The holistic approach was one that the respondents universally favoured, but which they felt they were unable to teach freely due to the pervasive nature of "traditional" approaches. Various reasons were given for this:

Teacher 1	I'm more in favour of a holistic curriculum, but in our school most of the curriculum is designed along traditional lines
Teacher 2	Traditional themes: that's how the older teachers like it
Teacher 3	I think we should focus on skills, but most curriculums are designed around themes, because that's how they used to have it in the National Curriculum

The teachers were expressing a perceived lack of authority and responsibility in relation to being able to decide what was 'on the curriculum'. The teachers appeared to feel that this relationship was not one that they "owned", but was passed on to them. This "passing on" may be in terms of official discourses through the ORF (such as the national curriculum or examination specifications) or through the PRF or more local discourses through the legacy of a particular department. Either way it appears to leave the teachers feeling incapacitated to take control of the geography curriculum that they are teaching.

12.7 Conclusions

In summary, from the data collected from this small scale pilot study, the definition of school geography in England appears to be influenced by a legacy of traditions—historical and social contexts pertinent to where school geography is taught. Such local definitions and distinctions can influence what it means to 'think geographically', and are therefore key to gaining a deeper understanding of the relationship between the academic discipline and the school subject. Therefore, in addition to Stengel's three factors, we also need to understand the legacy of how that relationship has been developed in the past, to reveal some of the forces that underpin how it is being interpreted by teachers. This finding suggests that there are likely to be significant regional as well as local differences in how school geography is understood.

These "legacy" contexts can act as a lens through which understandings of geography education research will need to be filtered before it can be enacted upon. In the data discussed above, how teachers feel about being able to interpret and create their local curriculum affect their actions. Legacy issues may be a significant barrier to teachers acting upon the findings of relevant research. This finding suggests that understanding these regional variations is important, but also that highlighting them is not sufficient: we also need to understand their impact on teachers, and the extent to which they can enable or restrict teachers making curriculum changes.

The pilot study seems to suggest that expanding research to other geographical locations would be valuable. Looking at the English context alone would seem to indicate that local factors have a significant impact on how geography is understood and defined, but also in how teachers feel enabled to act and change the curriculum. These findings, along with the findings from the other pilot countries, have been used to create an online survey tool which will be used to expand this research further. The survey tool will seek to explore not just how geography is defined but also the significant factors in influencing that definition of the subject. Upon completion of the survey in the original pilot countries (and Sweden), we hope to expand our research to further international contexts.

Such an understanding of how geography is understood in international contexts will reveal the degree of significance to these factors. This will help geography educators develop a nuanced understanding of the factors that can influence the local creation of the geography curriculum. Such a detailed understanding can then reveal important influences as to why thinking geographically differs from location to location. This pilot study suggests that these factors are not just related to inherent regional differences in the discipline itself, but the factors that influence its recontextualisation: some of which are local and influenced by the individual scale of the school.

References

Alexander, R. J. (2001). *Culture and pedagogy: International comparisons in primary education*. Oxford: Blackwell.

Bernstein, B. (1977). *Class, codes and control. Vol. 3. Towards a theory of educational transmissions* (2nd ed.). London: London: Routledge and Kegan Paul.

Bernstein, B. (1990). *The structuring of pedagogic discourse*. London: London: Routledge.

Butt, G., & Lambert, D. (2014). International perspectives on the future of geography education: an analysis of national curricula and standards. *International Research in Geographical and Environmental Education, 23*(1), 1–12. doi:10.1080/10382046.2013.858402.

Firth, R. (2012). Disordering the coalition government's 'new' approach to curriculum design and knowledge: The matter of the discipline. *Geography, 97*(2), 86–94.

Goodson, I. (1987). *School subjects and curriculum change: Studies in curriculum history* (2nd ed.). London: Falmer Press.

Maton, K. (2014). *Knowledge and knowers: Towards a realist sociology of education*. London: Routledge.

Morgan, J. (2011). Knowledge and the school geography curriculum: A rough guide for teachers. *Teaching Geography, 36*(3), 90–92.

Rawling, E. (2001). *Changing the subject: The impact of national policy on school geography 1980–2000*. Sheffield: Geographical Association.

Roberts, M. (2014). Powerful knowledge and geographical education. *Curriculum Journal, 25*(2), 187–209. doi:10.1080/09585176.2014.894481.

Seow, T. (2013). *The subject conceptions and practice of pre-service geography teachers in Singapore*. Unpublished Ph.D. Thesis, Institute of Education, London.

Smith, J. P., & Girod, M. (2003). John Dewey & psychologizing the subject-matter: big ideas, ambitious teaching, and teacher education. *Teaching and Teacher Education, 19*(3), 295–307. doi:10.1016/s0742-051x(03)00016-7.

Stengel, B. S. (1997). 'Academic discipline' and 'school subject': Contestable curricular concepts. *Journal of Curriculum Studies, 29*(5), 585–602. doi:10.1080/002202797183928.

Uhlenwinkel, A. (2012). *Reason and representation—The difficulties of avoiding maps without meaning*. Paper presented at 32nd International Geographical Union Congress, on August 2012 at Cologne, Germany.

Yates, L. (in press). Content selection as a curriculum issue: What, specifically, constitutes Physics as 'powerful knowledge'?

Young, M. F. D., Lambert, D., Roberts, C. R., & Roberts, M. D. (2014). *Knowledge and the future school: Curriculum and social justice*. London: Bloomsbury Academic.

Chapter 13
Getting Back to Basics: Is the Knowledge of School Geography Powerful in Chile?

Victor Salinas-Silva, Andoni Arenas-Martija and Laura Ramírez-Lira

13.1 Introduction

One of the reasons why powerful knowledge has become such a provocative concept is the underlying question of the extent to which school subjects have a role to play in current society. This is not just a utilitarian query but a gentle reminder that once in a while it is necessary to stop and reflect about the social conventions that inform teachers' practices.

Questioning geography's powerful knowledge has come at a bad time for school geography in Chile, as the 2013 curriculum changes have dismantled the subject between year 7 and year 10. In Chile History and Geography have traditionally been taught together (by the same teacher) but with independence of each other. This was the case until the reforms in the 1990s when interdisciplinary approaches were implemented and the contents were brought together with social sciences. Through this process, geography and the social sciences have become more subordinate to history on the school curriculum. Even though geography still has an explicit presence in the primary curriculum, continuity and progression of learning has been broken. Such a move could indicate that there is little that is considered powerful in the subject after all. But as the discussion on what is going to happen to geography in year 11 and 12 has not yet started, it has become increasingly important to recognise the social value of geography.

In recent years, curriculum changes tended to be a constant in Chile. However, Brooks (2010) suggests that, when faced with new curriculum policy documents, teachers tend to revert back to their tried and tested conceptions of the subject and how it should be taught (see for example Roberts 1995). It may be the case that teachers do not completely endorse the new guidelines that are prescribed from the central government.

V. Salinas-Silva (✉) · A. Arenas-Martija · L. Ramírez-Lira
Pontificia Universidad Católica de Valparaíso, Valparaíso, Chile
e-mail: victor.salinas@ucv.cl

C. Brooks et al. (eds.), *The Power of Geographical Thinking*,
International Perspectives on Geographical Education,
DOI 10.1007/978-3-319-49986-4_13

Teachers' conceptions of the subject are important because they define what is actually taught, and could provide an insight on what knowledge is finally selected in schools. But, that will not be sufficient to indicate if that knowledge is powerful. Young's (2008) provocation to consider what is powerful about geographical knowledge requires an examination of the connection between school content and academic disciplines. The relationship between these two dimensions has been examined by Stengel (1997) who argued that it is not always the case that these two separate spaces are in dialogue, or communicate knowledge with the same codes.

Although Young argues for specialised knowledge from a curriculum perspective, he does consider the question of how it is 'pedagogised', that is how it is 'paced, selected and sequenced for different groups of learners' (2011, p. 151). He argues that this requires an understanding of the differences between school and every day knowledge, knowledge domains and, specialist disciplinary knowledge (e.g. physics or history) and pedagogised knowledge (school physics or school history, for different groups of learners)' (Young 2011, p. 151). When exploring the epistemic relationship between disciplines and school subjects, it is necessary to point out that this relationship is not always direct (Stengel 1997). In the context of this study, the relationship is understood from the perspective of the teacher. The results shed light on teachers' notions of subject knowledge (Lambert and Jones 2013), and how it is related to the power of geographical thinking.

13.2 Powerful Knowledge: Lost in Translation?

The debate about Young's (2011) powerful knowledge is concerned with getting back to knowledge—what is the purpose of school for students and how students can move beyond the everyday knowledge that operates outside of school. In contradiction to the relativist-constructivist perspective, it is recognised that schools provide a point of access to specific, organised knowledge that is not commonly found in society. It is accepted that knowledge is socially constructed (Berger and Luckmann 1968) yet it is also understood that in order to teach something, that knowledge needs to be organised (Young 2008), which entails a high level of specialisation:

> Powerful knowledge refers to what knowledge can do – for example, whether it provides reliable explanations or new ways of thinking about the world … powerful knowledge in modern societies … is, increasingly, specialist knowledge (Young 2011, p. 150).

This specialist knowledge has been developed in schools in subjects that have traditionally been structured along the same lines as their academic counterpart (Stengel 1997). However the relationship between the two dimensions is not always direct and could be shaped by the type of link that exists in each school culture (with that culture acting as a mediating element).

The first mediating element stems from the paradigms that shape Chilean education in general. Social realism, as described by Young, challenges the politics that have allowed constructivism to become firmly established in Chilean educational discourse. A cognitive perspective prevails which has focused on the mental processes students develop in order to complete a task (Coll 1993). This has been the predominant way of understanding learning and has permeated teaching practices through national assessments and programmes guided by the paradigm of the effective school (Barber 2007). In this respect, the idea of powerful knowledge can be misinterpreted as a disproportionate defence of the scientific disciplines as a form of specialist knowledge, in a context where it is perceived that traditional or "hard knowledge" (Morin 2009) should be given less importance or be supplemented by skills that are of direct use in students' lives (Monereo 1997).

The second mediating element stems from what the public considers to be geography and moreover how geography has been understood by laypersons (Garrido 2005, 2013). In this respect, 'powerful' involves providing 'access to concepts which enable young people to move beyond their experience in ways that would not be open to them in their families and communities' (Young 2011, p. 145). Although this argument tends to tip the balance in favour of subject knowledge over everyday knowledge as suggested by Roberts (2014), this issue is not a clear-cut dichotomy. It is argued, for example, that map reading is context-dependent knowledge as it is developed to solve specific problems in everyday life (Young 2011). This skill can be acquired in different spaces other than in school. But, it is also a culturally acquired skill (Monereo 1997) which is developed differently from one country to another and with a level of specialisation that requires both geographical conventions and tacit knowledge on what a map is or what it is for. In any case, school plays a role in defining that knowledge, but it is also responding to a cultural requirement.

Young highlights specialised knowledge as a core element of powerful knowledge, generally defined through the discipline (Young 2008; Young and Muller 2014). However, Young's proposal does not fully address or provide an explanation for the evolution of knowledge, i.e. the way knowledge is constructed and therefore stabilised as a field of specialisation (Beck 2013). In his later work on the sociology of professions (Young and Muller 2014, p. 15) he acknowledges that the way a field of practice is 'codified, standardised or institutionalised will shape the form that professional knowledge in different fields takes'. The institution in which the knowledge is used will eventually remodel its understanding. This is an important issue that could delimit if powerful knowledge is defined through what happens in school (Catling and Martin 2011), or solely by the discipline. It should be analysed on a case by case basis, considering actors, traditions and paradigms affecting a particular set of knowledge.

Much of the aforementioned is mediated by different visions about how geography should be taught, which brings a third mediating element into play. Although subject content or knowledge is part of the curriculum, this does not necessarily mean that these will be borne out in the classroom (Rawling 2001). Faced with changes to the curriculum, teachers tend to revert to their preconceived or 'tried and tested' ideas about the subject, its contents and how these should be taught (Brooks 2010; Roberts 1995). In Chile, studies such as those by Araya (2013) and Arenas (2010) have approached this from the perspective of the didactics of geography, proposing that different ways of teaching and thinking about geography are informed by the objects studied and the epistemology constructed around the discipline.

In Chile this field of research is still being developed at the level of educational theory and is commonly understood as a problem of didactics, as the study of how to teach a discipline rather than just the resources employed in a lesson.[1] It involves issues around the relationship between pedagogy, curriculum making and subject expertise. However, according to Miranda (2012) it is symptomatic in its development as a field. There is still confusion in distinguishing geographical thinking, as originating from the geographical sciences, from spatial thinking, which is not organised in a prescribed way and is commonly used by lay people (Gardner 2008).

School content is commonly understood through the concepts or narratives it employs (Brooks 2013). Nevertheless, the methods developed by each academic discipline in terms of the way in which information is gathered and processed are part of the teaching of the school subject. At a pedagogical level in Chile, this is described using the term 'procedures'. According to Pozo and Postigo (2000) this reflects the need for students to know the processes used by each academic discipline to construct knowledge and to understand that these are not limited to the direct application of techniques and methodologies. A better understanding of the process can help to define teaching stages and build scaffolding that supports students' learning.

Finally, part of Young's argument is based on the relationship between powerful knowledge and the knowledge of the powerful. This idea can shed light on the recent debate about Chilean education given the high levels of social inequality and socio-economic segregation in the Chilean school system. From a curricular perspective, this notion has primarily been developed by critical pedagogy in Chile, echoing the studies by Giroux (1990) about the social function of schools as an instrument of control and Freire's (2005) pedagogy of the oppressed. Together with critical geography, the sociological perspective of these two fields challenges the cognitional hypotheses with which constructivism has been understood in Chile.

---

[1]Didactics is dedicated to the study of teacher-content-student relationship in the learning and teaching process of a particular school subject, like Mathematics (Biehler 1994), Natural Sciences (Adúriz-Bravo and Izquierdo 2002) and Geography (Monereo 1997). The way the concept is understood in the English-speaking tradition is different from the way it is understood in Spanish. While English-speaking traditions relate the concept to behaviourist practices, Spanish traditions consider didactics as the pedagogy, the method and practice of teaching. In some cases it is organised according to specific academic disciplines, with a scientific approach (Tochon 1999).

Didactics is understood as an epistemological consideration of knowledge and the teaching/learning processes of scientific knowledge (Benhamla 2012). It focuses on teaching practices and the pedagogical discourse generated about these practices (Galisson 1986), which in turn generates teaching strategies for a better understanding of a particular discipline (Academia de Lille 2006). It also focuses on the what and on the how of subject content and how it should be taught to reach certain educational goals, considering that what is taught is not a mechanical repetition of scientific knowledge given as certain and valid (Picardo 2005), but specific reconstruction of this knowledge for the school and its' students (Academia de Lille 2006).

There are different facets to the development of geography education in Chile and these have put a strain on the relationships that are required in order to maintain the knowledge structure that gives access to the powerful knowledge as outlined by Young. On a theoretical level one can assume there are certain mediating elements that aid understanding of these facets but also distance them from their original idea. This suggests that in order to effectively understand the power of geographical knowledge, it is necessary to try to understand the way in which geography as a school subject is interpreted and understood within this system, what influences this vision and why it is considered necessary for students to learn it.

13.3 Methods

Qualitative and empirical approaches were used (Schwandt 1998) to try to understand the relationship teachers make between their understanding of geography and what is actually taught. This considers the way the subject is structured and how its importance is determined, and reveals the scope of what can be understood as powerful knowledge in geography in the Chilean school context.

The study was conducted in parallel with researchers from other countries. The focus was to take into consideration the views of teachers in order to characterise geography and the way it is taught in each respective country. The approach was to allow teachers to generalise their responses rather than focusing solely on their own conceptions.

13.3.1 Data Collection

An instrumental and collective case study was the principal method used (Stake 1998) with an emphasis on qualitative data for researching the teaching of social sciences (Seale et al. 2004) as the school subject in the country includes: geography, history, economy and civics.

Five teachers were identified from networks working in state schools. Each teachers had a minimum of 3 years experience teaching History and Geography. These teachers were working in different types of primary and secondary schools in Chile's Valparaíso region: schools following a vocational curriculum, schools with a traditional humanities and science-based curriculum and rural and urban schools, as indicated in Table 13.1.

Data was collected through semi-structured Interviews (Taylor and Bogdan 1986). The data was recorded using field notes and audio recordings. Interviews were conducted between December 2014 and January 2015 and results were analysed between March and May 2015.

Table 13.1 Characterisation of teachers in the sample

Teacher	Teacher qualification	School type	School management	School funding	Curriculum	Students' age
Francisco	Secondary teacher	Urban	Public	Public	Science and Humanities	13–18
Javier	Primary teacher	Rural	Public	Public	Composite classes	6–12
Diego	Primary teacher	Rural	Public	Public	Composite classes	9–12
Silvia	Secondary teacher	Urban	Public	Public	Vocational	12–18
Pablo	Secondary teacher	Urban	Private	Public	Science and Humanities	17–18

13.3.2 Data Analysis

The data was analysed using content analysis and triangulation methods. Explicative and structuring content analysis (Flick 2007) based on emergent categories (Miles and Huberman 1994) were used (Table 13.2).

Triangulation (Lincoln and Guba 1985) was used for the in-depth analysis. These results are discussed on the basis of the work by Stengel (1997) and her analysis of the way school subjects relate to their academic disciplines. The focus was on how this relationship manifests itself in teachers' understanding and the interpretation of their academic discipline in the school context. This will make it possible to evaluate Young's proposal in terms of how geography is viewed and its perceived importance as a school subject.

Table 13.2 Emerging categories

Research question	Emergent categories
From a teacher's perspective, what is geography?	The characteristics of geography; What is studied in geography; geography for change; and limitations of geography
As a teacher, how do you think geography is understood in Chile?	Reduction to a thematic area; knowledge and subject exist in isolation; an iconic product i.e. maps; the study of the geographic characteristics of our country; and our understanding of geography is shaped by our individual experiences
What are the main influences on school geography?	Meaningful learning and behavioural practices; from descriptive learning to analytical learning; and geographical localisation as a basic skill

13.4 Findings

13.4.1 What Is Geography?

13.4.1.1 The Nature of Geography

This category refers to the academic area (i.e. science, humanities, arts) to which geography is attributed. Two of the respondents, Silvia and Pablo, felt geography was a science as it meets the requirement of having an object (geographical space) and a method (according to Pablo: cartography).

> For me geography is a science – it is a science insofar as it has a method. This method has specific stages, starting with observation, description, and everything is observed is integrated with everything observed. Geography as a science has an object of study, which in this case would be geographical space (Silvia, interview 4).

It was unclear whether Francisco, another respondent, considered geography to be a science or a humanities discipline. At first he said it was a humanities discipline as etymologically, he stated—it is not a science, but he later contradicted himself by saying that the aim of all of the sciences, including geography, should be able to explain the world.

13.4.1.2 Divisions Within Geography

Pablo and Silvia explicitly referred to the way geography is divided into different areas of study. Francisco implicitly acknowledged these divisions by using concepts such as 'physical geography' and 'population geography'.

Pablo made a distinction between physical and human geography, noting physical geography tends to be an exact science and studies events beyond human control, whereas human geography is a social science involving the study of things controlled by humans. He recognised there are many subdivisions within these divisions, but did not mention these during the interview.

Silvia referred to the different "types" of geography: physical, environmental, economic, urban, population and demographic (these last two overlap considerably), however she did not make the broader distinction between physical and human geography.

13.4.1.3 What Is Studied in Geography?

This category does not refer exclusively to geography's *object* of study, as only Pablo and Silvia mentioned this concept. It is instead a more generic approximation of what it is that geography tries to describe, explain and understand, according to the teachers. Two of the interviewees referred to geographical space as the object of study and the other three described geography as the study of the relationship between people and their environment.

Of the interviewees that considered geography to be the study of the human-environment relationship, Francisco and Javier both described a bi-directional relationship, although with some differences. Francisco referred to the environment as the "surroundings" (Francisco) which included the spatial territory and everything above (or beneath) it. However Javier described two relationships: the human-natural coexistence (a bi-directional relationship) and the human-natural-spatial relationship (with an emphasis on the local space), as evidenced by the following extract:

> So one part is nature with its purer forms and the way man, or man with nature, relates to this space. (…) I directly associate geography with territory, in which the inhabitant, the human being, the person, coexists with his natural environment (Javier, interview 2).

Diego referred to a unidirectional relationship in which humans engage with their surroundings through their activities and the way they use land and natural resources.

Pablo and Silvia said geography is the study of geographical space. The following statement helps to understand Pablo's view of what is studied in geography:

> [Geography] is characterised by having an object of study, which is the territory: its dynamics, its components and its interactions between territorial components (…) Geography has been characterised as a science of territory, of the relationship between territorial elements and the way in which humans inhabit, exploit, transform and degrade this territory (Pablo, interview 5).

Pablo sees geographical space as a territory, an "inhabited space". He defines inhabited space as a space with a direct or indirect human presence, because he also described geography as the study of physical phenomena that are not controlled by humans (but do affect them). As the previous quote shows, the object of study—the territory—includes all of the components, dynamics and interactions that exist and occur within it. Pablo's view coincides with Francisco and Javier's view in terms of a bi-directional relationship between humans and the territory, because by transforming and degrading the territory, humans can affect and be affected by it.

Silvia also mentions this relationship but not in response to the question about what is studied in geography. She refers to this relationship in terms of why geography should be studied, as the following quote shows:

> (…) so this subject is really important insofar as we know understanding how the earth functions with its systems with all of these things here [referring to the types of geography] (I might have missed out a few) we can have a better relationship with the geographical space, we can look after it (Silvia, interview 4).

According to Silvia, geographical space can be studied on its own in order to build a better relationship with it, or it can be studied in terms of the relationships between humans and geographical space, as some of the "types" of geography she mentioned previously.

13.4.1.4 Geography for Social Change

All of the interviewees said, either explicitly or implicitly, that geography plays a part in bringing about change in our society and environment. As Javier said: "So, the more we understand the coexistence or relationship between humans and nature, the more likely we are to generate an alternative social development."

According to Francisco, geography provides tools to make decisions about how people relate to their surroundings. Javier said geography allows him to coexist with nature and understand it. Diego argued geography enables his students to value their territory and make better use of the land. Pablo did not make any explicit assertions regarding this issue, however the examples he mentioned revealed the way in which geography can address social problems in Chile. Silvia said geography enables people to build a better relationship with their space which goes beyond sustainability, as evidenced by the following quote taken from her response to the question 'What is studied in geography?':

> (…) we can have a better relationship with geographical space, we can look after it. And we could even think beyond a concept of sustainability and beyond economic resources. When I talk about economic resources, I am already ascribing a value to water, fish; I am ascribing a value to everything. I even think it has to transcend this concept. For me this is why geography is important and this justifies studying and teaching geography (Silvia).

In Silvia's opinion, geography can shift the anthropocentric perspective of concepts like sustainability and resources towards a vision in which humans see themselves (in Silvia's words) as "guests on planet Earth".

13.4.1.5 Boundaries of Geography

Four of the interviewees referred to the overlap between geography and other disciplines or said its boundaries are blurred.

Francisco mentioned this with reference to how geography should be taught, saying it is easier to explain phenomena by making links with other subjects, as happens in the other sciences, given all of these subjects attempt to make sense of the world. Javier felt it was absurd to divide any kind of knowledge into sections and therefore there are no actual boundaries to geography or knowledge. According to Silvia, geography is a multidisciplinary science because in studying geographical space one must make links to many other disciplines. Pablo felt the recent move towards a more in-depth study in geography threatens to turn the subject into nothing more than an auxiliary science:

> I also think, although its field is very wide, it has gradually been reduced to the conceptual vision of what geography does, which has also given it breadth (…) I see it as an auxiliary science to other sciences and also as a science in itself has its own purpose (Pablo).

13.4.2 How Is Geography Understood in Chile?

13.4.2.1 Reduction to a Thematic Area

According to one of the interviewees (Francisco), people have a reductive view of geography in terms of its contents and themes. They see geography as a series of specific traditional "branches": physical geography and regional geography in an administrative-territorial sense. Reference to climate change could refer to a more complex geographical problem.

> In Chile, people see geography as physical geography, they never relate it to population dynamics, the enrichment or impoverishment of culture and the evolution of language (…) Geography is not just geography, it's not just about the landscape and volcanoes and rivers etc., it's more than that. So this is how I try to explain things (…) But if we think about geography, people think about regions, climates, some might think about climate change but I think the last thing people would think about (and this is a well-known fact) is the phenomenon of urbanisation and the rural world and population dynamics (Francisco).

The reduction of school geography to these thematic areas has been instrumental in transforming the subject into a specific form of knowledge: physical geography is seen as the primary, formal and fundamental aspect of any geographical study; while regional geography is seen as the territorialisation of space for the purposes of administration and use. However these have become the only possible forms of geography; curriculum designers, teachers and above all students are unaware of the thematic, theoretical and methodological advances have occurred in contemporary geography.

13.4.2.2 An Insular Knowledge and Subject

According to one of the other teachers (Javier), school geography is not only suffering from a diminished curriculum, it is also isolated from other school subjects and other sciences. The professional, scientific and academic spheres of geography are also seen as detached from other forms of knowledge and professional fields. According to Javier these views are erroneous and affect the way geography is taught in schools:

> In fact I teach the study programmes in reverse. In geography we are learning about the earth's layers and hydrology. Our topic this month is water. So taking water as our starting point we are studying all the subtopics. But to do this I am teaching the final unit first, the hydrosphere, the earth's layers, how they work, and by doing so we are emphasising the processes of geography (Javier).

To rebuild this missing link, school geography needs to address the segmentation of the curriculum (earth's layers, natural resources, hydrosphere) by identifying key topics that integrate with other subjects but with a specific focus on the way geography deals with these topics (such as water) while also making links to the real life issues affecting communities, schools and students (such as drought).

Geography as an academic discipline and as a school subject is segmented and disconnected. As an academic discipline that could and should explain reality it is isolated from a broader context; while as a school subject it is isolated from other subjects. According to Javier, the relational/integral dimension of geography as an academic discipline and a school subject has been reduced.

13.4.2.3 An Iconic Product: Maps

Two of the teachers (Pablo and Silvia) based on their experiences with students, said students feel learning geography is about making maps.

> …Based on conversations I've had with my students, I can say their previous experiences with geography have centred on the products geography can give you, for example maps. For them, doing a geography assignment means working with maps (Silvia).

If formal education focuses on specific products there is no opportunity or demand to question or reformulate these products and actions, they become static and permanent. Therefore geography becomes: "making maps".

13.4.2.4 Chile's Geographical Characteristics

Three teachers (Diego, Pablo, Silvia) referred to the geographical characteristics people associate with Chile. People express these using stereotypical images, such as Chile being "a long and thin" country:

> We always see Chile as a strip of land, there is always this cliché of Chile as a long, thin strip of land. But we are a country with a multitude of different landscapes and terrains are more than just long and thin (Diego).

People recognise (or not) certain geographical characteristics of Chile. This geographical identity is based on a set of attributes that are seen as specific and unique, so much so they become clichés. These attributes are based on the country's physical geography and geographical features. Being long and thin are the geographical characteristics of Chile. By ignoring other geographical features and ideas the country's image is restricted.

13.4.2.5 Our Understanding of Geography Is Shaped by Our Individual Experiences

Diego, Pablo and Silvia spoke about how individual experiences have shaped their ideas about geography: direct family experience such as through a father who treks; school experience which was restricted to map reading; and academic experiences guided by geography teachers. For one of them:

> … there is no awareness geography has a wider horizon and it can affect you personally, geography can improve your life, for example through public policies (Pablo).

Previous research (Roberts 2014) has shown how personal experiences of space are relevant to the explanations and meanings given to geography and the reasons for teaching geography. This experience linked to the direct, diverse and intense contact between teachers and places, may be relevant due to their physical or human characteristics. However, this enriched experience is not always accurate or is expressed in the classroom as a criterion for improving the positioning of geography (Arenas 2010).

13.4.3 What Are the Main Influences on School Geography?

13.4.3.1 Between Meaningful Learning and Behavioural Practices

Although there is a strong influence within the system towards fostering a constructivist approach in the classroom and a general acceptance of this approach among teachers, when asked about delivering the geography curriculum in schools, some of the teachers mentioned behavioural notions which, paradoxically, echo outmoded visions on the subject.

> For some aspects of geography, the best way to show it is in the 'geography laboratory' i.e. through field work (Francisco).

When Francisco refers to fieldwork as a specific methodology for teaching geography, he is describing a learning approach to "show" students aspects of geography. Teachers frequently say geography can only be taught in the field, or at least, this is the most effective way. Therefore the purpose of fieldwork is to expose students to a specific stimulus. You need to see the Andes to fully grasp the concept of mountains; otherwise you will have a reduced understanding.

The meaning of the word "showing" is rooted in the idea of transferring knowledge (Marzano and Pickering 2005). This tends to favour classroom activities or skills-based learning objectives that are focused on the recovery of information (Pozo 2011).

> I think people don't really understand what geography is, I think there is a lot of ignorance about the subject; for most people geography is what they learnt at school: how much water is there in the Loa River, how long is it, why is it the longest river in Chile?…ehh! What countries border Chile? and that's all…nobody realises geography has a wider horizon and it can affect you personally, geography can improve your life, for example through public policies" (Pablo).

This was most apparent to the teachers when teaching physical geography. When the interviewees talked about other colleagues, they suggested on the whole geography is dealt with in a descriptive way (see Monereo 1997).

> I think on a general, academic and school level, the conceptual change from a descriptive geography to a more analytical one has been understood. But in practise we haven't got away from the descriptive, most teachers continue, or maybe we continue, teaching geographical descriptions (Pablo).

The teachers tended to trace this problem back to their initial training. Four of the interviewees admitted they and their colleagues have at some point felt ill-equipped to teach geography. Although Pablo and Silvia recognised this is a generalisation, they said their confidence in their subject knowledge and training—compared to other colleagues—has had a positive impact on their teaching.

13.4.3.2 From Descriptive Learning to Analytical Learning

Even in this scenario, three of the interviewees referred to school geography from a conceptual perspective; using concepts as umbrella terms to guide their teaching or applying them directly to teaching geography. What Javier said reflects the former:

> How do we educate [the students] so they make sense of the territory from a social justice perspective and how can we tackle climate change (…) and how does this impact on the local area? (Javier).

Javier, Pablo and Silvia, tend to deliver the geography curriculum in a more conceptual way as opposed to a narrative or factual way. This means they are more flexible about what geography can do. Rather than categorising geography into topics, they organise it into broader areas or problems (such as in the case of Javier).

13.4.3.3 Geographical Localisation as a Basic Skill

A third difficulty can be viewed in terms of the skills students develop (Pozo and Postigo 2000). According to the interviewees, geography reinforces students' orientation skills. To understand geography you need to know where things are located. It would seem the teachers spend a lot of lesson time developing students' orientation skills.

Silvia noted geographical orientation is also seen as a way of making up for students' lack of cultural capital (Bourdieu and Passeron 2013), particularly for more disadvantaged students. Silvia explains "*they would point to Bolivia on the map and say that's where New York is!*" According to Silvia this is an example of the external pressures faced by teachers. Although teachers see these pressures as factors limiting their work, they also feel it is their professional duty to build a fairer society.

In this context, ensuring the progression of skills (Eggen and Kauchak 2005) is a significant challenge for teachers and involves creating a sort of baseline knowledge. Only when this is in place can they begin developing other skills that are more complex for students. Silvia notes a large proportion of Years 7 and 8 are spent developing these orientation skills and only in Years 10 and 11 can she start doing investigative activities to develop students' analytical and research skills.

13.5 Discussion

Young (2008) asks two questions that are fundamental to education: 'what is worthwhile knowledge?' and 'what should we teach?' (p. 15). But how is knowledge defined? Young claims it is the product of a process of specialisation (Young 2011). This type of knowledge can be likened to Scheffler's (1991) concept of "stores of truth" discussed by Stengel (1997, p. 586). Stores of truth equate to all of the specialist knowledge accumulated by academic disciplines which go into making the curriculum (as a continuous relationship between academic disciplines and school subjects (Stengel 1997). The specialist knowledge of academic disciplines is therefore powerful knowledge as it explains reality in a consistent manner and cannot be acquired from everyday experience. So is geographical knowledge the same as powerful knowledge according to teachers? Is it relevant to teaching and learning?

In terms of being a powerful knowledge, when Silvia and Pablo define geography as a science they are already making a distinction between the knowledge geography produces and the knowledge acquired through experience—as geographical knowledge is delineated by an object and a method. Meanwhile when Francisco says geography tries to explain the world, he is not necessarily making a distinction between geographical knowledge and the everyday knowledge acquired through experience (because there is no indication that he sees geography as a science), but he is attributing one of the qualities of powerful knowledge to geography, i.e. its authority over other types of knowledge. Although the other interviewees did not explicitly allude to geographical knowledge as originating from a science, or from a specialist discipline, by defining the field of study they were implicitly assigning it the task of explaining a portion of reality. All of the interviewees, when giving examples of what they taught in geography, mentioned concepts and forms of visualising reality as part of geographical knowledge and as forms of knowledge which are not acquired from everyday life.

In terms of the relevance of geographical knowledge, all of the interviewees felt it has a role to play in bringing about social and environmental change. This makes geographical knowledge particularly important in the context of social inequality and climate change at a national and global level. Therefore giving students access to geographical knowledge and acquiring the tools and vision it can provide, is relevant.

Although not all of the interviewees alluded to geographical knowledge as powerful knowledge, they all considered it to be relevant for society and therefore, relevant to teaching and learning.

Stengel (1997) however, expressed a concern about any curriculum that is based on the division of knowledge—as this detracts from the moral nature of teachers' work. The teachers in this sample echoed Stengel's concern. For example, Francisco and Javier suggested geography should not be taught in isolation but alongside other subjects. Javier even argued that any division of knowledge is an illusion, because knowledge is a single entity. Like Stengel (1997), Pablo suggested it is this lack of clearly defined borders between geography and other disciplines that tends to render geography an auxiliary science.

Geographical knowledge is restricted by at least two of the traditional pathways for discussing school geography: initial teacher training and the official curriculum, which has resulted in an acutely discontinuous relationship (Stengel 1997) between the school subject and disciplinary knowledge. According to the teachers, most people's view of geography and Chile's geographical attributes is partial, limited and of little social significance; Young would describe this as creating limited geographical knowledge which had little power. This limited view prevents people from being able to explain social phenomena. The interviewees stated they have been able to overcome these limitations and establish a more continuous relationship between core knowledge and the school subject (Stengel 1997), developing their own discourse about teaching geography.

It is hard to define the actual learning that is accessed by students through school geography as it tends to vary according to the culture which prevails in each school. It can range from what we could call the creation of a *little geographe*r, i.e. a student who is able to use the basic knowledge and methodological tools of geography, to a student who can *think geographically*, with a mental and conceptual framework which enables the student to view real life spatial problems through the lens of geography.

The existence of a behavioural and in some cases descriptive approach to teaching geography indicates there have been a series of mechanisms that have made these practices resilient to the shift towards a constructivist curriculum that has prevailed in Chile since 1996. Terms such as "meaningful learning" or "the students' context" would appear to indicate the way teachers understand pedagogy. But it is not clear if they are referring directly to the hypotheses of Vygotsky (1964) or Piaget (1970) on this issue, or if their interpretation (Collins and Evans 2002) of subject's knowledge position would be mistaken with the type of understanding Young (2008) has. Similarly, the resilience of pockets of knowledge (Jasanoff 2003) in school geography is an interesting area to study in order to understand how effectively a specific type of knowledge is instilled, updated and can endure in teachers' subject expertise (Brooks 2010).

13.6 Conclusions

This chapter has partially shown that according to a sample of teachers, school geography in Chile contains powerful knowledge in terms of its usefulness for understanding the world (beyond the everyday) and taking actions. To some extent geography education is also seen as a means of accessing social justice in Chile.

Based on Stengel's ideas about the relationship between academic disciplines and school subjects, it is possible to assess the relevance of Young's hypotheses in the particular contexts of these Chilean teachers. In doing so, findings shows a continuous relationship persists in the understanding of geography, however in terms of public perception and school geography this is not so clear. Although this does not call into question the idea of powerful knowledge, it indicates a necessity of making a review of the way specialisation in the subject is conceived in the case of teachers for a context where geographical training is not fully developed.

The link between the hypotheses of Young and Stengel poses an interesting question for future research: In what ways do teachers' experiences, and the context in which they work, condition their conceptualisations about geography and how it should be taught? During this study findings provided some evidence of a relationship between these factors, suggesting teachers in rural areas develop a much more contextual vision of geography, based on elements and phenomena that are relevant in their local area. Meanwhile teachers in urban areas have a more global vision of geography that encompasses a multifaceted area of study.

Similarly, teachers working in rural areas did not refer to academic geography and simply applied concepts that were most relevant to their local context. On the contrary, teachers in urban areas based their definitions of geography on their ideas of academic geography and sought to reflect these ideas in their teaching.

These elements will possibly provide evidence to contribute to the discussion on the social value of geography as a school subject considering how powerful it is to foster learning in different school contexts, which in turn could provide a positive argument about its importance and presence in the school curriculum.

References

Academia de Lille. (2006). Léxico de términos pedagógicos generalmente utilizados en el mundo educativo y de la enseñanza/Artes plásticas. Septiembre, Lille, Francia. http://artsplastiques.discipline.ac-lille.fr/

Adúriz-Bravo, A., & Izquierdo, M. (2002). Acerca de la didáctica de las ciencias como disciplina autónoma. *Revista electrónica de enseñanza de las ciencias, 1*(3), 130–140.

Araya, F. (2013). Estrategias docentes para el desarrollo de habilidades de pensamiento espacial en República Dominicana y Chile. *Revista geográfica de Valparaíso, 47*, 27–41.

Arenas, A. (2010). *Las estrategias, actividades y tareas de enseñanza en el aprendizaje comprensivo de la historia y las ciencias sociales escolares. Casos chilenos. Tesis doctoral.* Madrid: Universidad de Alcalá.

Barber, M. (2007). *How the world's best-performing school systems come out on top*. London: McKinsey & Company.

Beck, J. (2013). Powerful knowledge, esoteric knowledge, curriculum knowledge. *Cambridge Journal of Education, 43*(2), 177–193 (2013). doi:10.1080/0305764X.2013.767880

Benhamla, Z. (2012). De la pédagogie à la didactologie: histoire d'une discipline ou histoire de concepts? *Synergies Algérie* n° *15*, pp 13–23.

Berger, P., & Luckmann, T. (1968). *La construcción social de la realidad*. Buenos Aires: Amorrortu.

Biehler, R. (1994). Didactics of mathematics as a scientific discipline (Vol. 13). Berlin: Springer Science & Business Media.

Bourdieu, P., & Passeron, J. C. (2013). *Los herederos: los estudiantes y la cultura*. Buenos Aires: Siglo Veintiuno Editores.

Brooks, C. (2010 Autumn). Why geography teachers' subject expertise matters. *Geography, 95*(3), 143–148.

Brooks, C. (2013). How do we understand conceptual development in school geography? In D. Lambert & M. Jones (Eds.), *Debates in geography education* (pp. 75–88). London: Routledge.
Catling, S., & Martin, F. (2011). Contesting powerful knowledge: The primary geographical curriculum as an articulation between academic and children's (ethno-) geographies. *The Curriculum Journal, 22*(3), 317–335 (2011). doi:10.1080/09585176.2011.601624
Coll, C. (1993). *El constructivismo en el aula*. Barcelona: Graó.
Collins, H. M., & Evans, R. J. (2002). The third wave of science studies: Studies of expertise and experience. *Social Studies of Sciences, 32*(2), 235–296.
Eggen, P., & Kauchak, D. (2005). *Estrategias docentes*. México D.F.: FCE.
Flick, U. (2007). *Designing qualitative research*. London: SAGE.
Freire, P. (2005). *Pedagogía del oprimido*. México D.F.: Siglo XXI.
Galisson, R. (1986). Eloge de la didactologie/didactique des langues et des cultures. *Etudes de Linguistique Appliquée* N°64 (pp. 39–54). Paris: Didier Érudition.
Gardner, H. (2008). *La mente no escolarizada*. Buenos Aires: Argentina.
Garrido, M. (2005). El espacio por aprender, el mismo que enseñar: Las urgencias de la educación geográfica. *Cad. Cedes, Campinas, 25*(66), 137–163.
Garrido, M. (2013). The place where waters murmur: Taught and learned andean space. *Review of International Geographical Education Online RIGEO, 3*(1), 26–55.
Giroux, H. (1990). *Los profesores como intelectuales*. Barcelona: Paidos.
Jasanoff, S. (2003). Breaking the waves in science studies: comment on H.M. Collins and Roberts Evans, 'The third wave of science studies'. *Social Studies of Science, 2003*(33), 389.
Lambert, D., & Jones, M. (2013). *Debates on Geography Education*. London: Routledge.
Lincoln, Y., & Guba, E. (1985). *Naturalistic inquiry*. London: SAGE.
Marzano, R., & Pickering, D. (2005). *Dimensiones del aprendizaje*. Mexico D.F.: ITESO.
Miles, M., & Huberman, A. M. (1994). *Qualitative data analysis*. London: SAGE.
Miranda, P. (2012). La educación geográfica en chile: desde su aparición en el currículum escolar en el siglo xix hasta los ajustes curriculares de 2010. *Revista Anekumene, 4*, 51–71.
Monereo, C. (1997). *Las estrategias de aprendizaje*. Barcelona: Edebé.
Morin, E. (2009). *Introducción al pensamiento complejo*. Barcelona: Gedisa.
Piaget, J. (1970). *La construcción de lo real en el niño*. Buenos Aires: Proteo.
Picardo, O. (2005). *Diccionario Enciclopédico de Ciencias de la Educación*. San Salvador: Centro de Investigación Educativa, Colegio García Flamenco.
Pozo, J. (2011). *Nuevas formas de pensar la enseñanza y el aprendizaje*. Barcelona: Graó.
Pozo, J., & Postigo, Y. (2000). *Los procedimientos como contenidos escolares*. Barcelona: Edebé.
Rawling, E. (2001). *Changing the subject: The impact of national policy on school geography 1980–2000*. Sheffield: Geographical Association.
Roberts, M. (1995). Interpretations of the geography national curriculum: A common curriculum for all? *Journal of Curriculum Studies, 27*(2), 187–205.
Roberts, M. (2014). Powerful knowledge and geographical education. *The Curriculum Journal, 25* (2), 187–209. doi:10.1080/09585176.2014.894481
Scheffler, I. (1991). Basic mathematical skills. In I. Scheffler (Ed.), *In praise of the cognitive emotions and other essays in the philosophy of education*. New York: Routledge.
Schwandt, T. (1998). Constructivist, interpretivist approaches to human inquiry. In N. Denzin & Y. Lincoln (Eds.), *Handbook of qualitative research* (pp. 118–137). London: Sage.
Seale, C., Gobo, G., Gubrium, J., & Silverman, D. (2004). *Qualitative Research Practice*. London: SAGE.
Stake, R. M. (1998). *Investigación con estudios de casos*. Madrid: Morata.
Stengel, B. (1997). 'Academic discipline' and 'school subject': contestable curricular concepts. *Journal of Curricular Studies, 29*(5), 585–602. doi:10.1080/002202797183928
Taylor, S. J., & Bogdan, B. (1986). *Introducción a los métodos cualitativos de investigación*. Madrid: Paidós.
Tochon, F. V. (1999). Semiotic foundations for building the new didactics: an introduction to the prototype features of the discipline. *Instructional Science, 27*(1–2), 9–32.

Vygotsky, L. S. (1964). *Pensamiento y lenguaje: teoría del desarrollo cultural de las funciones psíquicas*. Buenos Aires: Lautaro.
Young, M. F. D. (2008). Bringing knowledge back in: From social constructivism to social realism in the sociology of education. Oxon: Routledge.
Young, M. F. D. (2011). What are schools for? *Educaçâo Sociedade & Culturas, 32*, 145–155.
Young M. & Muller J. (2014) *Knowledge, Expertise & the Professions*. London: Routledge

Chapter 14
Teaching to Develop Geographical Thinking

Felisbela Martins

14.1 Introduction

The aim of this chapter is to explore the theoretical approaches that can inform the training of geography teachers so that they are able to support the development of geographical thinking.

Geographical thinking is increasingly important. The economic, social and cultural changes in the global world make the school a meeting place of different cultures. Students need to acquire social and cognitive skills, as well as preparing themselves to face the day-to-day issues and decisions. The role of geography in the school context should allow students to discover, at various scales, the global world in which we live, and to give them the skills they need to operate in such a world. In other words, students should be able to understand the world so that they can act and react in their environment. The aim of this study was to explore the methodology that teachers should focus on in order to promote geographical thinking which can foster young people's ability to think and act in the spatial environment where they live and "to become self-fulfilled and competent individuals, informed and aware citizens and critical and creative 'knowledge workers'" (Lambert 2009, p. 17).

To promote geographical thinking in this way requires a conceptual and methodological change in teacher education, such as a focus on curriculum concepts that relate to the local context, and an emphasis on the two-way relationships of teachers as managers of the curriculum development process and of their professional qualifications. In this chapter, we discuss the concepts and methods of

F. Martins (✉)
Faculty of Arts and Humanities, CEGOT—Centre of Studies on Geography and Spatial Planning, University of Porto, Porto, Portugal
e-mail: felisbela.martins@gmail.com

C. Brooks et al. (eds.), *The Power of Geographical Thinking*,
International Perspectives on Geographical Education,
DOI 10.1007/978-3-319-49986-4_14

teaching Geography in schools, as well as the development of teacher practices regarding which geography should be taught, and how it should be taught. This discussion is with a view to how we can develop geographical thinking.

14.2 The Curriculum and Its Design by Teachers

Today knowledge is regarded as extensive and dynamic, and is widely accessible in a variety of formats and open to a range of audiences. The school, however, is still very much a privileged institution for conveying knowledge in a systematic way to young people, and the curriculum "is the defining principle at the centre of the school" (Roldão 1999, p. 26). In this context, the school can be seen as an important institution that does not just convey knowledge but also enables people to understand how to use that knowledge.

In a society strongly influenced and dependent on knowledge, education continues to be a valuable resource for social inclusion, contributing to the personal development and skills of every individual—imperatives that are essential requirements to their autonomy and social life—and to the integration and social use of knowledge. But the question still remains as to how to achieve this.

Zabalza (2003, p. 17) defines the school curriculum, as being

> all sorts of arrangements and processes that each society sets up so that the right of children and young people to education and personal and educational development becomes effective through the schools. Thus 'curriculum' herein refers to what is done at schools, that is, the response given by schools/and/or through schools in relation to the right of children and young people to education

Roldão (1999, p. 26), elaborates that: "the set of lessons learned that, because they are socially necessary in a given period in time and context, must be guaranteed and organised by the school". This places the school centrally as a key context for defining the education that takes place, and makes it even more important that schools question what they want to teach, to whom and for what purpose. Underlying these issues are fundamental concepts on the design and purpose of education. It is not reasonable to think that in a multicultural society the curriculum should be focused on the supposed compliance with universal programmes for all school students.

Rethinking the curriculum means re-engaging with questions about what is meant to be learnt in today's world, and what aspects of the past should be included. According to Roldão (1999, p. 36), the curriculum must integrate

> a range of different *learning contents* including: reference knowledge areas without which access to knowledge cannot be put into operation; activation and consolidation of autonomous knowledge building; command of tools to access knowledge in its various forms; development of social attitudes and skills; mechanisms for individual development and improvement of quality of life (emphasis in the original).

In this dimension, the curriculum must no longer be understood merely as a programme to be taught to students; instead, it must be understood as a *corpus* that incorporates a set of measures and actions which may be perceived differently by different people. It can no longer be what Formosinho (1991, p. 262) called a "uniform curriculum—ready to wear, one-size-fits-all" without the need for it to be thought about, as argued by Roldão (1999, p. 38), "according to a number of drivers of change". Roldão continues that these drivers include

> the need for *different* curriculum proposals based on *common targets*; the focus on the acquisition of desirable levels of *skills* in the fields covered by school education; the anchoring of curricular practices on *significant* references and contexts for all those attending school; the reconstruction of the curriculum as a *specific project* of each school, adapted by its actors and managers, replacing the *standard discourse* by the *situatedness* discourse (*ibidem*, emphasis in the original).

By taking this approach and thinking about the curriculum in this way (and not just as a space where one can manage and make decisions) implies new roles for both schools and teachers. As stated by Zabalza (1992, p. 47), "it is to understand the curriculum as an area for decision-making in which, based on the programme and by programming, the school community, at school level, and the teacher, at class level, articulate their respective frameworks for action". It also implies, however, that students should no longer be the recipients of the conveyed information, but agents producing their own knowledge—capable of living in society today and in the future.

This perspective places the school as a site of decision-making, able to adapt the National Curriculum to the local context, and where teachers can act as agents capable of playing an active role and becoming curriculum designers and managers. This perspective also takes a different approach on how the school curriculum is defined. The curriculum

> should not be perceived merely as a set of subjects to be taught in pre-made information packages, or even as a set of highly structured school programmes that allow students to develop the required skills, but rather as a set of intentions, means and activities that enable students to actively participate in building their own knowledge and developing themselves in a positive way, both social and socially. (Leite 2003, p. 90).

While this is one way of ensuring that everyone will learn more, this equality requires the curriculum to be thought about differently, and in particular through consideration of how subjects are represented within it.

This perspective also places teachers in a different role—as agents involved in managing the implementation of the curriculum, who regard their work as not being limited to the classroom, but who take it beyond the classroom to the broader school, community and society context. Teachers who question the education taught per se, who question the actual act of teaching and thinking about the curriculum, and whose role is that of "a researcher of their own teaching practice" (Stenhouse 1987, p. 195).

14.3 Teachers as Curriculum Development Managers Versus Their Qualifying Professional Development

To regard teachers as curriculum designers presumes that they play a role in the management of curriculum development processes and are mediators between the National Curriculum and students' prior knowledge. Teachers must not only have a good command of what they teach, but also understand the purposes that underpin their teaching.

Teachers are responsible for the transition between the formal curriculum and the real curriculum, and act as curriculum development agents. Teachers who think about their teaching practice (Schön 1998) must decide on learning priorities for each class and the particular aspect of knowledge (disciplinary or otherwise) to be addressed, whilst bearing in mind the students' learning methods and their specific experiences and needs in terms of joint organisation (per class) of learning to be achieved in the class (Roldão 1999). By thinking about their actions and teaching practice, teachers make individual decisions on how to organise learning and their curricular learning materials, produce and implement working methods for each class, organise each class or group, and assess learning improvements achieved in view of the work strategies used. From this point of view, teachers perform a number of procedures and have to adapt the contents and instructions received from the central administration. Consequently, a top-down approach is not possible nor desirable. This reflects a view of teachers as: "professionals who must play an active role in formulating both the purposes and objectives of their work, and the means to achieve them" (Zeichner 1993, p. 16), who are capable of "taking active responsibility for raising serious questions about what to teach, how to teach, and which broader goals they are striving for" (Giroux 1997, p. 163). This chimes with Zeichner's descriptive of reflexive teachers (1993, p. 16) or, as Giroux (1997, p. 63), puts it, "teachers as transformative intellectuals".

Thus, in a changing social context, the professional growth of teachers is paramount. Being a good teacher means presenting and explaining the subject well, conveying the programme contents so that everyone can access disciplinary knowledge. However, in the past, the curriculum was considered as a set of previously identified contents and objectives, to be taught to students in the same way using, preferably, textbook(s). Accordingly, teachers simply consumed and implemented the curriculum, which was meant to comply with central government instructions. From this angle, the primary aim of teacher training was to "enable teachers to acquire a number of skills and knowledge needed to implement the guidelines established under the prescribed curriculum" (Flores 2000, p. 151). Teacher training in this context therefore involved building a training model centred on the subjects in order to privilege and encourage scientific and pedagogical-didactic knowledge. It was important for teacher training to be based on the scientific knowledge of the area of expertise and of techniques that would allow teachers to convey such knowledge. This was the training model in use and promoted in Portugal in the 1970s and 1980s. At that time, emphasis was given to a model focused on training for knowledge, in which the relationship between the teacher, the contents to be taught and the students in the classroom prevailed.

Today, however, there is a need to reconsider a teacher training model "that allows the acquisition and reconstruction of skills, knowledge and provisions to reinterpret the curriculum, requiring teachers to be more involved in their construction and development process" (Flores 2000, p. 151). Such a training module would "familiarises future teachers with the profession, allowing them to develop their curricular mentality (being aware of the system in which they are found and of the role, or roles, they play or could play in this system)" and prepare them to be "co-authors of the curriculum, rather than mere receptors" (Leite 2005, p. 138). This approach also encourages constant reflection, analysis and critique about teachers' own practice (Garcia 1999, p. 131) that extends beyond the relationship characterised by being "linear and hierarchical between scientific knowledge and its practice" (Flores 2000, p. 156). According to Leite (2005, p. 129), this implies "a professional socialisation that broadens the scope of disciplinary knowledge to its social uses", emphasising "the enhancement of cooperation practices during the training period(s) and the encouragement of experiences that promote the development of a participatory awareness".

This form of training deviates from the technicist model of teacher qualification, and allows teachers to be aware of their role as educational agents that reflect on and learn about their own practices, but also about a new pedagogical attitude and a new stance in education. In other words, critically reflective teachers who are capable of self-analysis and developing an autonomous approach to professional development.

14.4 Teachers and the Teaching Practice

There is an argument that every teacher must embrace a teaching-learning process that promotes knowledge, taking into account the specificity of the subject and of the environment in which learning takes place. This implies having a professional attitude, in which teachers reflect on their own practice (Alarcão 1996), organising their classroom work in an active and participatory way—as opposed to a model wherein the teacher appears as a mere conveyor of knowledge, which is still so prevalent in Portugal.

According to Roldão, the principle that "he who knows, teaches" (2009, p. 42), is outdated, as it is based on the notion that "the teacher is essentially a conveyor of knowledge and it is enough to know well in order to communicate well" (Estrela and Estrela 1977, p. 27). Today, however, teachers are required to know much more than simply how to teach. As stated by Roldão (2009, p. 42) "Those who know how to teach, teach, because they know what they are teaching and how to do it, to whom and what for". Thus, teachers should always take into account the importance of articulating what they learned as theory and their own practice. Teachers must master theoretical—scientific knowledge, but also methodological teaching skills.

In their practice, teachers are responsible for the progress of each student's learning. To do this, teachers have to take into consideration the student's prior learning and how to organise a number of learning opportunities to meet the intended objectives. In order for good learning to occur, the teacher must have a facilitating and emancipatory role as regards knowledge, leading their students into thinking, helping them to find ways of reflecting and questioning. However, being a facilitator also means that teachers need to give every student the opportunity to interact, raise questions and share knowledge and experiences. This classroom interaction is intended to build knowledge, develop competences, and, at the same time, is a learning opportunity for the teacher.

For Vygotsky (1998), a person's learning process should, above all, be considered as a development of cognitive structures. The formation and development of cognitive structures depends on how a person perceives the psychological aspects of the personal, physical and social world—it is a personal construction. Accordingly, the teacher then has to recognise the personal worlds of their students and to develop activities and tasks so that new learning is meaningful to them. Building on what the student already knows, that is, what is embedded in their cognitive structure, creates a link between the new contents and the student's previous knowledge, in an active and personal process. It is thus up to the teacher to know how to properly manage the learning process according to various parameters, one of which is the cognitive development of students. This means creating learning situations that allow interaction at various levels, because it is through this interaction that students will build knowledge (Vygotsky 1998). This observation has a powerful message for developing geographical thinking, as it means that it is only when teachers discover students' personal geographies that they can help them to come to deeper geographical understanding.

14.5 Training Geography Teachers with a View to the Development of Geographical Thinking

In my own institution, we have been using these principles to develop an initial teacher training programme which will enable students to become familiar with the geographical and pedagogical discourse, and to learn about the theoretical concepts that guide curriculum development. Our aim is to highlight the role of the teacher as a researcher and promoter of autonomy, who is able to engage with the planning and assessment of the teaching-learning process. Beginning teachers are required to master effective and original ways of understanding didactics, so that they may become effective geography teachers.

When students first come to school, they already have some geographical knowledge. The programme highlights that teachers have to put into practice geography education through cognitive and metacognitive strategies so that students can be aware of their own knowledge. In this constructivist context, the planning of the teaching-learning process should be based on these pre-concepts and prior learning, always bearing in mind that new concepts will only be appropriated by the students if they are meaningful, and if they understand them in the light of their own conceptual frameworks. This is central to our model of how they can develop geographical thinking.

However, this approach also requires a different approach to developing the theoretical and practical tools teachers need—such as for creating lesson plans. The planning model developed is based on conceptual planning, the starting point of which is the students' previous knowledge. Students are then guided along successive stages until they solve particular problem situations using new geographical knowledge and concepts. Along these lines, planning is "dynamic, open, flexible and incomplete [...] a planning process in action" (Braga 2004, p. 29). Thus planning is part of the curriculum development process, and teachers have new roles as curriculum designers, creating, implementing and assessing the educational processes. To embrace this new approach, teachers must be more involved and aware of the school community in general, and their students in particular. Such awareness is important for teachers to understand students' prior understanding and to help them build their knowledge and geographical thinking based on these concepts. In the

> constructivist approach planning involves creating stimulating environments that give scope to activities that cannot automatically be foreseeable and that, (...), cater for the various situations and the students' different starting points. This presupposes providing activities whose contents can become significant, helping students develop *learning to learn* skills (Braga 2004, p. 27).

As part of the teacher training programme teachers in a school context prepare various types of planning according to the different moments of the teaching-learning process. This is explored in one module, the Didactic Planning Unit (PUD), which allows teachers to develop each teaching unit and to organise their educational practice, adapting the teaching-learning process to the students' needs. The PUD is: "a group of contents and associated skills perceived as a logical unit" (Arends 2008, p. 60). López (1992, p. 79), in turn, defines a didactic unit as "a work unit relating to an articulated, complete teaching-learning process". Our students are encouraged to design a PUD with all the structuring elements (initial assessment, thematic, procedural and attitudinal contents, educational situations, and assessment), revolving around a focal point. They are encouraged to make didactic decisions in order to organise sequences of teaching-learning articulated around an Aggregating Educational Situation (SEA).

Future teachers are expected to carry out an integrated management of the Geography curriculum, that is, to design the teaching of the discipline "in a non-fragmented, non-centred way, specifically addressing context x, y or z, without a linear sequence, lesson after lesson, concept after concept, etc." (Martins 2011, p. 237). Our purpose is to "show that it is possible to design a plan in an integrated way that can be used in practical terms, in order to break away from routines and achieve new practices, so that new ways of developing geographical education can be embodied" (*ibidem*). This approach is more than "implementing" the national curriculum in a technical way, but is aimed to develop geographical thinking around structured educational situations. Although the focus is primarily on the cognitive development of students, it is useful to consider Coll's (1986) typology about the diversity of contents, namely: conceptual, procedural, and attitudinal. Such contents may include facts, principles, concepts, procedures and attitudes that can be applied to any geographical theme.

For example, this approach can be illustrated through the theme "Mobility of the population". The objectives proposed by the National Curriculum are: understand the causes and consequences of migration; understand and know the main international migration cycles, and know and understand migration in Portugal with regard to space and time. When designing the Didactic Unit, we started with the Aggregating Educational Situation "World Summit on Mobility" under the theme "The World in Motion". Following this Summit, the International Report on Migrations-2015 is expected to be available online, as the result of the students' work (Fig. 14.1).

What followed was a simulation of the Summit in which students play various roles. The following thematic key contents are identified, including: mobility, migration, types of migrations, causes of migrations, consequences of migrations, migration flows. The next stage is to identify geographical procedures, such as: use of geographical vocabulary, location of places and regions, collection and organisation of information, reading and interpreting graphs, maps, texts, images, systematisation of information collected, data processing and interpretation, and speaking and writing. In addition, attitudinal contents are defined: interest, commitment, responsibility, accuracy, reflection, autonomy, critical spirit, solidarity.

The students then focussed on eight different educational situations based on real cases, at national, European and world scales. Each situation was led by a congressman who represented each of the roles, such as "Portuguese Prime Minister, Captain of the Spanish hospital-ship Esperanza del Mar, Director of a Youth Shelter for Migrants from Nagykaniza, Hungary, President of the Christian Aid Organisation, Mayor of Bangalore, CNN Journalist in Sudan—Darfur, Minister of Health of Portugal, two geographers from the University of Porto". Everyone gave a speech, using maps, images, and statistical data to address themes such as World Population Mobility, the phenomena of migrations from the North of Africa and the Middle-East to Europe, movements of people because of conflicts, natural disasters and large development projects, new migration flows in regions with skilled labour employability, travel within regions devastated by wars, and finally, the large migratory flows in Portugal since the mid-19th century and migratory phenomena today. During the activity, some students have more important roles than others: some collect the information, record and interpret it, and in the end prepare a final report—the International Report on Migrations-2015, which is made available online.

The aim of this approach is to focus on developing an understanding of real, social, spatial, dynamic problems and their application (Hugonie 1989). Cachinho (2000) argues that these problems are all the more significant the closer they are to the students' lives and to how they affect the society in which we live. Through this approach, students can reflect on major social and environmental problems, by

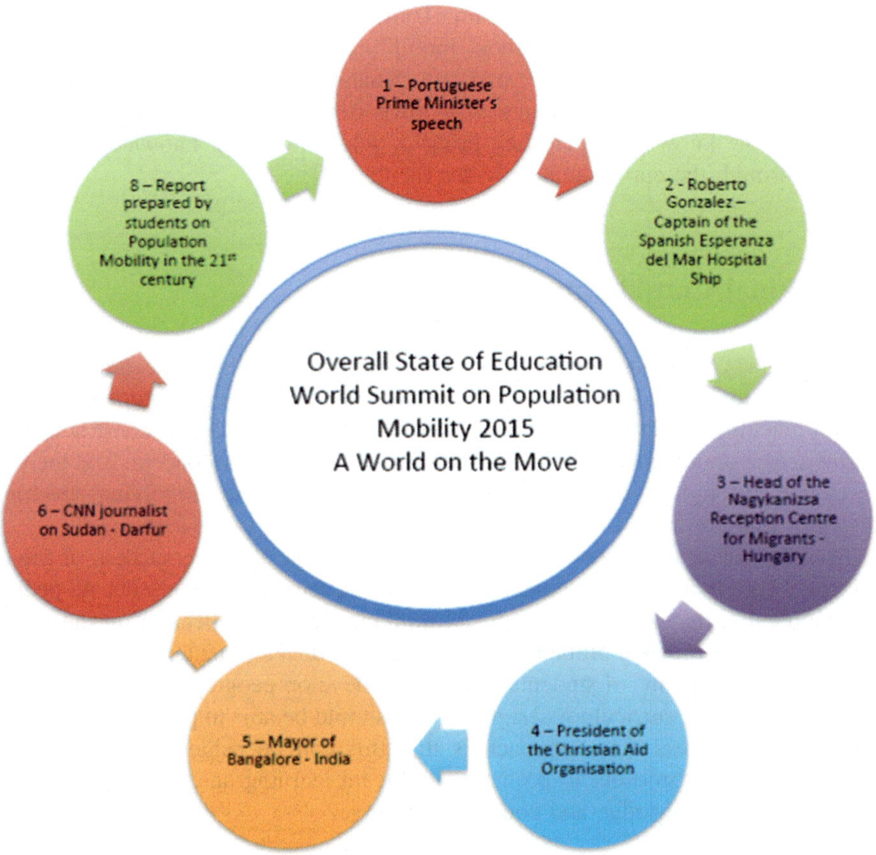

Fig. 14.1 Scheme of the Didactic Unit "Mobility of the population" under the aggregating educational situation "A World on the Move"

teaching students to formulate questions, and to pose critical questions about the information so as to address such problems competently. The teaching of school geography can therefore use geographical thinking to help students to become critical thinkers and to build their own geographical knowledge (Callai 1999). In this case, the theme under study "Population Mobility", must be thoroughly analyzed in a systematic manner, including their elements and relations, comparing analyses at different scales and acknowledging that relations and spatial processes change according to changes in geographical scale (Lacoste 1980; Cachinho 2000).

The geographical thinking lies in the conceptualization exercises and in the comparison of problems at various scales, preparing students to know how to think about space and to conscientiously act in the environment in which they live. Moreover, the problems under study should be addressed actively, because it is important to allow students to address real problems, mobilizing their knowledge, procedures and geographical techniques.

After designing the Didactic Planning Unit, each teacher is able to put this into practice, making students think about mobility and migration at the level of the local, national and world space, and the dynamics and interrelations occurring therein. This means that students will have developed geographical thinking, combining the knowledge that they bring to school with the knowledge that the school, through the curriculum, provides for them.

14.6 Final Considerations

The National Curriculum needs to be designed at a local level, according to the school and to its students. As such, teachers need to be curriculum designers. Within teaching, teachers should be concerned with their students' construction of learning. Our approach is that this should start with their prior concepts and then be developed through a series of opportunities for learning, involving teaching strategies that are innovative, motivating and meaningful to students.

As a school subject, geography can contribute greatly to the success of future citizens. More than teaching geographical content, teachers should educate people geographically, developing geographical thinking. It is by reflecting on the major social and environmental problems, focusing on the development of the real, spatial and everyday problems of students that we can develop geographical thinking. For teachers to become curriculum designers, they should be able to plan the curriculum development process. Our approach is that this can be achieved by proposing problem situations in order to promote significant learning and, consequently, the construction of knowledge and conceptual change.

Acknowledgements This work was co-financed by the European Regional Development Fund (ERDF) through the COMPETE 2020—Operational Programme Competitiveness and Internationalization (POCI) and national funds by FCT under the POCI-01-0145-FEDER-006891 project (FCT Ref: UID/GEO/04084/2013).

References

Alarcão, I. (1996). *Formação reflexiva de professores – Estratégias de supervisão*. Porto: Porto Editora.

Arends, R.I. (2008). *Aprender a Ensinar* (7a. ed.). Madrid, SAU: Mcgraw Hill Interamericana de Espanha.

Braga, F. (Ed.). (2004). *Planificação: Novos papéis, novos modelos: dos projectos de planificação à planificação em projecto*. Porto: Edições ASA.

Cachinho, H. (2000). Geografia Escolar: Orientação Teórica e Praxis Didáctica. In *Inforgeo* nº 15. Educação Geográfica. Associação Portuguesa de Geógrafos.

Callai, H. (1999). A Geografia no ensino Médio. In *Terra Livre* nº 14. Associação dos Geógrafos Brasileiros.

Coll, C., Pozo, J., Sarabia, B., & Valles, E. (1996). *Los contenidos en la reforma Enseñanza y aprendizaje de conceptos, prodecimientos y actitudes* (2.a ed.). Madrid, S.A: Ediciones Santillana.

Estrela, M., & Estrela, A. (1977). *Perspectivas actuais sobre a formação de professores*. Lisboa: Editorial Estampa.

Flores, M. A. (2000). Currículo, formação e desenvolvimento profissional. In J. A. Pacheco (org.), *Políticas de Integração Curricular*. Porto: Porto Editora, Colecção Currículo, Políticas e Práticas, nº 3, pp.147–165.

Formosinho, J. (1991). Currículo Uniforme – pronto-a-vestrir de tamanho único. In F. Machado e F. Goncalves (1991), *Desenvolvimento Curricular – Problemas e perspectivas*. Porto: Edições ASA, pp. 262–267.

Garcia, C. M. (1999). *Formação de Professores. Para uma mudança educativa* Porto: Porto Editora, Colecção Ciências da Educação, Século XXI, nº2.

Giroux, H. A. (1997). *Os professores como intelectuais. Rumo a uma pedagogia crítica da aprendizagem*. Porto Alegre: Artmed Editora.

Hugonie, G. (1989). Enseigner la geographie actuelle dans les lycées. *L'éspace Geographique, 2*, 127–129.

Lacoste, Y. (1980). Les différrents niveaux d'analyse du raisonnement géographique. *Hérodote, 18*, 3–15.

Lambert, D. (2009). Geography in Education. Lost in the post? An inaugural professional lecture. Institute of Education University of London.

Leite, C. (2003). *Para uma escola curricularmente inteligente*. Porto: Edições ASA.

Leite, C. (2005). O Currículo e o exercício profissional docente face aos desafios sociais desta transição de século. In C. E. Ferraço (org.) (2005), *Cotidiano escolar, formação de professores (as) e currículo* Cortez Editora, Série Cultural, Memória e Currículo, nº6, pp. 128–139.

López, L. P. (1992). *Diseño de currículo en el aula: una propuesta de autoformación*. Madrid: Mare Nostrum.

Martins, M. F. S. (2011). *O Currículo Nacional de Geografia no Ensino Básico. Da concepção e do Prescrito à Confiruração do vivido*. Ph.D. thesis, Universidade do Porto, Porto, Portugal.

Roldão, M. C. (1999). *Gestão Curricular. Fundamentos e Práticas* Lisboa: Ministério da Educação, Departamento do Ensino Básico.

Roldão, M. (2009). *Estratégias de ensino: O saber e o agir do professor*. Vila Nova de Gaia: Fundação Manuel Leão.

Schön, D. (1998). *El Profesional reflexivo*: *cómo piensan los profesionales cuando actúan*. Barcelona: Paidós.

Stenhouse, L. (1987). Investigación y desarrollo del curriculum Madrid: Morata.

Vygotsky, L. (1998). *A formação social da mente: O desenvolvimento dos processos psicológicos superiores* (6a ed.). São Paulo: Martins Fontes.

Zabalza, M. (1992). *Planificações e desenvolvimento curricular na escola*. Porto: Edições Asa.

Zabalza, M. (2003). A construção do currículo: a diversidade numa escola para todos. In Ó. C. Sousa, M. M. Ricardo (org.), *Uma Escola com Sentido: o Currículo em Análise e Debate. Contextos, questões e perspectivas* Lisboa: Edições Universitárias Lusófonas.

Zeichner, K. (1993). *A formação Reflexiva de Professores: Ideias e Práticas* Lisboa: FPCEUL, Educa.

Chapter 15
English Geography Textbook Authors' Perspectives on Developing Pupils' Geographical Knowledge and Thinking

Simon Catling and Jongwon Lee

15.1 Introduction

Teaching geography in schools serves a variety of purposes, developing pupils' geographical knowledge and understanding about the world and fostering their curiosity and fascination with the Earth and its peoples, enabling them to consider the opportunities and challenges faced presently and for the future (Lambert and Morgan 2010; Scoffham 2010). In doing this, the underlying interest is to engage pupils' geographical thinking and reflection (G.A. 2009). Pupils' geographical thinking is aided by building up a body of information about the world and its people, such as knowing its basic 'layout' of continents and oceans and the locations of key features, places, regions and countries. It is enhanced through enabling pupils to appreciate the range of ways in which geography seeks to describe, analyse, explain and evaluate how the world works, from meeting the needs and desires of everyday life to the processes which create and support these and which act on and sustain the Earth's natural environments. Geographical thinking requires understanding from a local dimension to a global perspective. Thus, it is underpinned by the information about the processes which create and affect the world as we know it and by the underlying concepts which geographers use to understand the physical and human processes continuously at work (Martin 2006; Bonnett 2008; Morgan 2012; Rawding 2013).

There are a variety of ways in which geography may be taught and a variety of resources available to use. One resource frequently used around the world is the school geography textbook. It acts as a key source of geographical knowledge and

S. Catling (✉)
Oxford Brookes University, Oxford, UK
e-mail: sjcatling@brookes.ac.uk

J. Lee
Ewha Womans University, Seodaemun-gu, South Korea
e-mail: jongwonlee@ewha.ec.kr

C. Brooks et al. (eds.), *The Power of Geographical Thinking*,
International Perspectives on Geographical Education,
DOI 10.1007/978-3-319-49986-4_15

is used to encourage pupils' thinking through a geographical lens. A textbook is thought to provide structure for teaching geography, even, perhaps when used independently by pupils as the geography 'instructor'. A debate, initiated in England during the early 2010s, about the role and value of textbooks in its schools has focused largely on the textbooks of subjects other than geography (Maughan et al. 2015). The view promoted is that high quality textbooks have a positive effect on pupils' subject learning (Oates 2014; Gibb 2014, 2015). It is based on an analysis of textbooks from such nations as Singapore, the U.S.A., China and Finland. The high use of textbooks in countries such as these is contrasted with comparatively low textbook usage in England.[1] While there seems not to be support for a move in England to introduce government approved textbooks, as occurs in several nations, the argument is that high quality textbooks provide the basis for effective pupil learning in subjects at an introductory level in primary schools and in developing pupils' understanding of relevant content and expectations during their secondary schooling towards national examinations—though this is contested, not least because of the likely impact of highly effective teachers (Maughan et al. 2015). This debate is contentious (Alexander 2015; Apple 2014) in that concerns are raised about the ideological basis of national textbooks, the constraints they may well provide if linked only with a national curriculum and examination syllabuses, and their use by teachers whose understanding of the school subject and its teaching is either evolving or possibly quite limited. This critique does not undervalue textbooks as a useful, even valuable, resource which can help to broaden pupils' knowledge and understanding and to enhance their horizons; nor is it to say that textbooks might not be used effectively in teaching.

Textbooks are authored. In some countries panels are commissioned to oversee textbook production and provision, perhaps with team members writing the approved textbooks. In other nations core textbook content may be prescribed, with publishers tendering to produce them or by publishing textbooks for state approval. In many countries, such as England, the preparation and publishing of textbooks is a commercial decision made by independent publishers, though a number of these may well seek the endorsement, for instance, of examination boards. There has been a range of research into the nature and quality of geography textbooks, changes in their content and style, and their selection, purchase and use by schools in the UK and elsewhere (for example: Graves 2001; Hamann 2007; Hopkin 2001; Janko and Knecht 2013; Lambert 1998; Marsden 2001; Walford 1995). Yet what appears to have been of little interest is the perspectives of textbook authors on the writing of their textbooks, considering for instance why they write and what they intend to achieve. These are matters pertinent to debates about geographical knowledge and thinking.

---

[1]The low use of textbooks in English Schools may be due in part to critical perspectives about textbooks as pedagogical vehicles (Samuda 2005). In fact one textbook author of the current study noted that 'Textbooks are tainted in secondary schools. Teachers I met wear as a badge of honour that they don't use textbooks. Often the textbook has merely been replaced by PowerPoint [that is, visual and verbal slide presentations resourced from a website or a CD-ROM]' [S2].

15.2 A Question of Knowledge and Thinking in Geography Textbooks

Discussion of the role of knowledge in geographical education has emerged in recent years (Firth 2011, 2015; Morgan 2013; Young et al. 2014). This has been framed in the context of social realist perspectives on geography, arguing that there is something powerful about the knowledge which academic study of geography in school provides for younger people's insights into the world, through both its key concepts (such as, place, scale and interconnections) and the range of themes which geography studies (including population growth and migration, climate distribution and changes, urbanism, water quality and access, transportation, environmental hazards and their impact, and much more). The disciplinary context is seen as underpinning the school subject, enabling its role in school to take pupils beyond what is referred to as their 'everyday knowledge', which is identified as their geographical experiences in daily life. The essence of this debate is that the primacy of subject knowledge needs to be resurrected to be clearly evident in the school curriculum (Young 2008; Young and Muller 2016), not least in geography.

School textbooks are a traditional means for providing subject matter—indeed, subject knowledge—for school pupils. In geography, their role has been to set out facts about the world alongside introducing pupils to geography's themes and topics, together with information on and insights into the world's countries and continents. Fairgrieve (1937) noted that textbooks are of vital importance for geography teaching in primary classes, though he did not discuss the knowledge they might disseminate nor how they could best be used. Writing at around the same time, Scott (n.d. 20, 21) noted that textbooks should contain 'carefully selected important geographical facts', which, for instance, needed to connect with and appeal to elementary school pupils' 'interests and experiences' when first they encounter a geography textbook. He emphasised that textbooks should explain, for instance, why places, regions or industries had developed and were important locally, nationally or internationally. Pupils should develop their locational knowledge but neither indiscriminately nor over fully. Scott argued that a key role for geography textbooks was to enable pupils to build their knowledge base about their home nation and the world, though he made no reference to geographical thinking. Indeed, Fairgrieve (1937), while he discusses the content of secondary school geography teaching, says little about the role of secondary textbooks and barely mentions thinking geographically.

Little has been written directly about the role of geography textbooks in promoting geographical knowledge, understanding and thinking in subsequent decades. Long and Roberson (1966), for instance, discussed a number of features of textbooks, particularly in relation to their content, style and the balance of their use of a range of written and visual texts. They commented on concerns about whether textbooks are up-to-date and accurate, and discussed how teachers and pupils might use them, but they provided no considered discussion of the textbook's role in developing pupils' geographical understanding and thinking. Cox (1973) highlighted developments in the discipline of geography and proposed that school textbooks should reflect such developments. He promoted the importance of currency and accuracy but added that the subject's ideas and methods should underpin secondary textbook writing. He averred, also, that authors' understanding of the textbook as a pedagogical device was important. Young (1977), an experienced geography textbook author, expressed the view that factual information, through case studies, should enable pupils to develop understanding of geography's concepts and themes.

Lidstone (1992) noted that textbooks can take a more open approach to geographical knowledge and encourage geographical thinking through critique and application, though this appeared from his research not necessarily to be a consistent intention or active approach within geography textbooks. Alongside these matters, he acknowledged rising concerns about possible underlying ideological stances and the biases, stereotypes and selective emphases contained in geography textbooks. These are issues that affect how geographical knowledge has been and continues to be portrayed (Ahier 1988; Hamann 2007; Hopkin 2001; Morgan 2003; Norcia 2010; Peled-Elhanan 2012). Nonetheless, he argued that textbooks have a role in updating geography teachers' knowledge and understanding, as well as in promoting newer approaches to its teaching. Lambert (1998) acknowledged this role, though he raised concern about the extent to which textbooks might be sampled for homework tasks rather than read for understanding, thus acting inappropriately as a resource in geography teaching and learning. Waugh (2000), considering his own approach to geography textbook writing, indicated that he saw their role to be to provide a core of material upon which teachers could build. For him being up-to-date, relevant and factually accurate were important, as was enabling pupils to enjoy the subject and fostering their interest in it. He felt that his personal experiences of places, as an author, was vital (Freeman 2008). However, he had little to say about the nature and role of knowledge in geography textbooks, though he inferred that applied thinking was an important aspect in geographical learning.

In their historical analyses of geography textbooks Graves (2001) and Marsden (2001) discussed subject content matters, the impact of national curricula, pedagogical interests and the influences of publishers. They acknowledged the role of school geography textbooks as repositories of disciplinary perspectives and content knowledge for pupils—and teachers—and reflected on issues such as stereotyping and nationalism. They concluded that by the turn of the twenty-first century the focus in textbooks had moved towards balancing the provision of geographical knowledge and engaging pupils' in its application, with commensurate increases in the use of interactive exercises, though they noted that reproductive and comprehension tasks remained. Geography textbooks could serve knowledge enhancing roles for pupils and promote subject thinking through knowledge application. While not discussing knowledge matters, Freeman (2008, 6), who interviewed a small sample of geography educators, including textbook authors, concluded that they construct a 'Geographical World' for their pupil readers. In effect, she argued, textbook geography is a version of geography, a representation which, for instance, textbook authors create and which reflect their own sense of geography, the school contexts for which they write and a variety of external influences, such as government subject policies and accessible resources. Additionally, Freeman noted that authors took little account of pupils' geographical experiences referring to an element of 'disconnect' by authors with young people's geographies (Freeman 2008).

Almost all the studies referred to above, as well as others (for instance, Marsden 1988; Walford 1995; Westaway and Rawling 1998), were based on reviews and analyses of geography textbooks. As Wright (1996) noted, rarely have there been interviews with (see Young 1977), or articles written by, geography textbook authors (Waugh 2000). Indeed, there appears to have been little interest in the perspectives of authors on their writing of geography textbooks, examining for instance why they write their textbooks and what they intend to achieve, the pedagogic approaches they use and what influences their eventual product, be it an individual textbook or a series of books covering a primary or secondary age range. Indeed, in the wider context of textbook research, this focus appears not to have been of interest, even in recent years (Bruillard 2011; Chambliss and Calfee 1998; Fan 2011; Foster 2011; Fuchs 2011).

15.3 Investigating Geography Textbook Authors' Perspectives

Textbook authors are an important source to investigate since their role involves mediating aspects of the knowledge of their subject for teachers to draw upon and for pupils to learn about. In effect, they provide, even unintentionally, parameters for what is taught and learned, as Graves (2001) and Marsden (2001) note. Through their selection of content authors can influence not just the subject areas which pupils know but how they think in the subject context and their attitudes to the subject (Freeman 2008). In geographical education, geography textbooks may well influence directions in pupils' geographical thinking (Graves 2001). This study did not investigate the ways in which *pupils*' geographical thinking is influenced or their application of geographical thinking. This research focused on the intentions of school geography *textbook authors* by investigating their perspectives on their writing intentions and expertise (Lee and Catling 2016). Intentionally, it did not relate their views about their textbook writing to their published textbooks.[2] The research question in this analysis was: do authors intend that their textbooks influence and shape pupils' geographical thought? If they do, what do they take into account?

This investigation was undertaken to explore textbook authors' perceptions of the educational goals to which they hoped their textbooks might contribute, their valuing of geographical learning for pupils, its benefits and the experiences that textbooks provided, strategies they employed to create learning progression in their

[2]Maintaining textbook authors' anonymity was agreed to because to identify their textbooks would have indicated the participant 'subjects' in this research. It was required to ensure their participation and confidence in the study.

textbooks, and the knowledge and expertise they perceived they needed in order to write an effective textbook. Through these topics it sought to identify what was important to these authors in writing their geography textbooks. It was vital that they were not led but given open-ended questions to respond to. While it was not a direct question to the authors, one interest implicit in this research was whether they aimed to support pupils' geographical learning through engaging pupils' geographical experiences, given the heightened interest in children's and youth geographies (see, for instance: Spencer and Blades 2006; Freeman and Tranter 2011; Holt 2011; Griffiths 2013; Somerville and Green 2015).

The study focused on authors writing geography textbooks for primary and secondary schools in England which followed the English national curriculum and examination syllabuses. The study was small scale and interpretivist. Other than gathering some background factual data, the questionnaire used asked open-ended questions that required the personal reflective views of the authors. The use of a questionnaire is not without its constraints, such as asking follow-up prompt, clarification or probing questions. Nevertheless the participants were given the freedom to write the length of response they wished to and they had a two month period over which to compose their responses. Feedback about responding to the questionnaire indicated that authors used this time to reflect on the questions and to work through and rework their responses prior to the set deadline. A constant comparative analysis of the responses was undertaken to identify the threads, common ground and variations in the perspectives of the authors (Newby 2014; Robson 2011). Seven of the ten English textbook authors initially approached responded positively to a request to participate in the study, split three to four between primary [P] and secondary [S] geography textbook authors. Several perspectives were identified, grounded in the authors' descriptions and views. In this report numbers are rarely given for the views expressed because the number of participants is low and there was a high degree of common ground. Nevertheless, a variety of viewpoints are reflected about geographical knowledge and thinking, though there is much in common in the perspectives of these authors.

It is appreciated that this is a small scale study, from which it is not reasonable to make generalisations. Many of the authors who participated were known to the authors of this paper. While there have been many geography textbooks authored since the introduction of England's national curriculum in the late 1980s, there is not a large number of authors writing in the 2010s, chiefly because the merger of publishing companies has led to a reduction in textbook publishing, particularly for primary schools.

15.4 Authors' Intentions in Writing Geography Textbooks

In the 1990s and 2000s in England, primary and secondary geography textbooks focused heavily on covering the geography national curriculum requirements and examination syllabuses for five to eighteen year olds. This sample of authors, who wrote textbooks in the latter part of this period, viewed this positively. England's national curriculum created through its various iterations a body of geographical knowledge (D.E.S. 1991; D.f.E.E./Q.C.A. 1999; D.f.E 2013; Q.C.A. 2007) which provided a core of geography to interpret and provide materials to exemplify, enabling authors to be thoughtful and creative. It provided opportunities to extend and deepen the sense of geography which authors wished to promote. One secondary author was encouraged by the geography requirements to argue that a textbook could provide text and activities which develop 'geographical thinking accessibly and at the right level of challenge' through the 'accuracy and currency of representing people and places through material' [S3]. A primary author expressed this opportunity differently, arguing for underpinning purposes to geography textbook writing:

> The importance of underlying values and principles, especially a concern for the environment and social issues in providing a touchstone for the selection of content [P1].

The prime intention expressed across this sample of English geography textbook authors was to write textbooks which, in the words of another primary writer, would

> produce a very simple and manageable series of books that progressed through key national curriculum themes in a way which would appeal to the non-specialist primary teacher and pupils themselves [P3].

It was felt by the primary textbook authors that primary teachers, who are almost always non-specialists in geography and whose confidence in their teaching of the subject might be low, would find the series supportive and engaging, enabling them to teach geography rather than not do so or work in a poorly planned and uncoordinated way. A textbook provides structure and enables teachers to teach; it supports pupils' learning (Graves 2001; Maughan et al. 2015). This approach is similar, while differently focused, to that espoused by the author of secondary school examination textbooks, who stated that a textbook is there to help pupils 'get through exams, so I have concentrated on providing what will work/help in exams' [S1]. While this author assumed that geography specialist teachers would provide good quality teaching for fourteen to eighteen year old pupils, it was felt that textbooks could equally help such pupils in independent study particularly when revising for examinations.

This sample of primary and secondary textbook authors was clear that their purpose in writing their geography textbooks was as summarised by this author:

> Today textbooks do not simply provide information and knowledge for teachers; they also have to provide a package which helps with the presentation and teaching ideas for that knowledge to be delivered well [P2].

Yet what emerged from the analysis of their perspectives on their writing indicated that their intentions were more deeply focused. The clear aim was to develop pupils' knowledge of geography, and within this, inherently, their geographical thinking. Underpinning this was the authors' subject and pedagogical knowledge, which they stated they applied, whether within a single textbook or across a series of books, to develop pupils' geographical understanding progressively. This was not always straightforward to achieve, they stated, since publishers retain an evident say in what they are prepared to publish, their interest in sales being paramount. These matters are elaborated below, drawing at further points on the words of this study's authors.

The authors' reasons for writing geography textbooks were to achieve three things. First was to foster pupils' geographical knowledge and understanding in order to promote their geographical thinking. Second was to promote the subject. Third, they intended to engage pupils' interest in and fascination with geography to achieve reasons one and two. What lay behind these intentions was the authors' enthusiasm for geography. Whether stated explicitly or embedded in their rationales for writing their textbooks, they expressed the view that geography is the best subject 'for helping pupils make sense of the world, and how it is changing… It is rich, relevant, exciting, engaging, and with a strong human element. On any day of the year, world and national news provide a stimulus for geographical thinking' [S4]. These three intentions are now considered further.

15.5 To Develop Pupils' Geographical Knowledge and Thinking

Recognising that primary and some lower secondary school teachers can lack many or some aspects of geographical knowledge, it was clear to the authors that they needed to provide a sound knowledge base in their books, reflecting Fairgrieve's (1937) and Graves' (2001) arguments. They recognised that through their textbooks, whether used by inexperienced or knowledgeable teachers of geography, they provided a framework of understanding and knowledge, which might well update teachers' sense of the subject (Lidstone 1992; Marsden 2001), even perhaps creating a 'geographical world' (Freeman 2008). Developing this understanding of geography inevitably reflected the English national curriculum requirements of the time (D.f.E.E./Q.C.A. 1999; D.f.E. 2013; Q.C.A. 2007), including geography's key concepts, required places and themes, and skills:

> Key concepts such as place, space and interaction provide geographers with a framework to make sense of what they discover. Skills such as mapwork, fieldwork and enquiry play an important role in informing the different interpretations which emerge… What makes geography unique (is) the way that geographers synthesise information from different sources [P1].

The focus on geographical knowledge was interrelated with geographical explanations to foster geographical understanding and thinking:

> Develop a framework of knowledge and understanding of the world, including place knowledge, knowledge and understanding of important processes and systems in the natural and human world [S3].

Another author drew on the 'importance of geography' statement from the 2007 revision to the key stage 3 programme of study to emphasise the importance of enquiry-based teaching and learning in geography and to underpin the view that this approach enables pupils to develop and apply their geographical thinking:

> Geographical enquiry encourages questioning, investigation and critical thinking about issues affecting the world and people's lives, now and in the future (Q.C.A. 2007, 101).

The authors argued that the development of geographical understanding should be fostered through the use of real examples which 'illustrate how things happened, what places are really like and how people adapt to living there. I like to help children to get an idea of what it would be like to really be there' [P2] and 'project the 'stories' of real people in real places' [P3]. This view emphasised the point that geography is and should be about the 'real world', using case studies to help pupils develop awareness, information and understanding of a variety of places and topics, while exploring key geographical ideas, reinforcing the perspectives of Young (1977) and Waugh (2000). For instance, this should include understanding the variations and patterns across the Earth, how peoples' lives and experiences vary from 'place to place, why this happens, that inequalities can be very real and that knowledge and understanding of this is a social good' [P3]. This indicated that knowledge and explanation need to go hand-in-hand. Authors noted that 'real world' examples should be used at a range of scales, requiring pupils to apply their geographical thinking to different aspects of the world's geography.

Authors averred that geography enables pupils to identify and understand change, providing insight into the vibrancy and dynamism of places. Using pupils' own worlds—their geographies—is a vital way to provide connections, for instance through traffic and recycling issues with younger children and in looking at the effects of glacial melt and trade cycles with secondary pupils. In considering environmental change, particularly in relation to local concerns, pupils can be encouraged to consider the impacts they might have by understanding such matters and expressing their views and proposals in response to textbook case studies and tasks as well as through textbook generated investigations, which almost all authors advocated through fieldwork enquiries in extension activities.

One author emphasised that the role of textbooks in developing geographical knowledge and thinking was to raise issues and to engage even young pupils in considering controversial matters, helping them to begin to realise that this might not bring straightforward resolutions. The purpose in examining concerns about the world's changing climate events and patterns was to foster pupils' geographical thinking through developing understanding of the gathering and interpretation of scientific data, that this is debated and interpretations challenged, and about which individuals need to make up their own minds, reflecting Lidstone's (1992) argument.

> I chose to use a topic such as climate change to try and illustrate how such an important challenge is being viewed by people around the world. I explained how scientific data is being collected and how this is quite controversial in many people's eyes. Children often find not having a yes/no answer available quite unsatisfactory, so they have to learn how to manage so many of the questions which do not attract a straight answer [P2].

It was through such approaches in their textbook writing that these authors sought to promote geography as an informative, problematic and stimulating subject, which could provide information about the world, understanding of its processes and key concepts. They sought to involve pupils in recognising that in dealing with many aspects of the world's geography, from the local to a global scale, there are not necessarily straightforward explanations and solutions to the challenges which people face. In this way, the authors intended to present geography as a vital and challenging subject, linked with pupils' lives and connecting their lives to the wider world beyond their local experiences.

To stimulate interest in geography most authors argued that it was important to introduce geography 'in a lively and engaging way' [P1] such that it interested pupils 'in the world they are growing up into' [S3], and enabled them to enjoy their geographical studies. Encouraging their interest in local and more distant geography is in part what geography teaching is about. Developing this attitude can lead, it is contended, to pupils valuing what geography offers and helping them to make better sense of their everyday geographies and what is happening in the world. This leads pupils to be better informed and more able and confident to apply their geographical learning to thinking about the world around them and of which they read, watch and hear. To achieve this, authors set out to 'present materials as clearly and engagingly as possible—and of course accurately' [S4], even when geography textbooks are written 'to interpret a specified syllabus' [S3], as Long and Roberson (1966) pointed out. This might involve using 'a wide range of resources' and 'exciting and thought provoking images' [S2] to provide stimulating material to draw pupils into the topic or place. One author illustrated this point about writing for secondary pupils as follows,

> In this book, it was mainly through the 'wow' factor, e.g. a case study on abortions worldwide and the status of workers on waste tips, and one on the health and social status of Australia's aborigines [S1].

15.6 The Effect of Textbook Authors' Knowledge

15.6.1 Subject Knowledge

The development of geographical knowledge and thinking through textbook writing required, in these authors' eyes, that they were able to draw on their own strong subject knowledge. This meant that they needed to be up-to-date with the full range of the subject, as Waugh (2000) has argued. To maintain this authors' need to keep up with developments in the subject, as well as to seek material which would help their textbooks 'to be 'fresh' with new case studies and new and interesting approaches to teaching geographical ideas' [S3]. They felt that this required of them 'a deep interest in geography' [S1] as a subject, such that they would undertake research into the topics they had selected to identify new material to include (Cox 1973; Shulman 2004; Lee and Catling 2016).

For these geography textbook authors developing pupils' understanding of the key concepts of geography was essential to fostering their geographical knowledge and thinking (Cox 1973). Thus it was vital to foreground key concepts through the content and geographical skills they selected. As an example, a core aspect of geographical knowledge—in the sense of information pupils needed to grasp—was noted to be locational knowledge, in effect pupils building increasingly wider and more robust mental maps of their locality and the world, through which to relate new and updated information to support making sense of the various features, events, places, environments and geographical topics and themes they studied—a point made in the 1930s by Scott (n.d.). This required writing which used 'grade appropriate' everyday vocabulary and subject terminology as pupils moved through their schooling. Likewise authors planned to draw on multiple perspectives in their case studies of places, themes and issues to engage pupils in considering the complexity of the world and as essential to enable them to challenge bias and stereotyping (Lidstone 1992; Marsden 2001). Thus, authors' subject knowledge needed to be current and informed by issues about knowledge, selectivity and interpretation as much as to give vibrancy to the subject, to test accuracy and balance, as to support pupils' learning of the geographical elements required through the primary and secondary national curriculum and examination syllabuses.

15.6.2 Pedagogical Knowledge and Approaches

Authors felt it was essential to be clear about ways in which a textbook could be the basis for effective geographical learning—how it might stimulate pupils' willingness to engage with the content, skills and ideas developed through the text. This required knowledge by the authors of what might be termed 'textbook pedagogy', as Long and Roberson (1966) inferred, that is, the organisation of pages and double-page spreads, the balance of visual images, data charts and written text on a page, the selection of language, vocabulary, pictures, maps and tables, and the sequencing of the material. Textbook spreads usually included tasks for the pupils, these needed to be appropriately pitched (as did the text) and to be 'varied, interesting, challenging and adaptable for teachers so they can ensure they are relevant to their pupils in their local environment' [S2]. Indeed, this author made the point that activities must challenge pupils in order to engage them in and support their geographical thinking.

A core concern was being skilful in conveying geographical knowledge in the confined space of a page or double page. Allying this with the challenge, as an author, of using 'interesting case studies, a variety of contemporary issues, a range of resources, [and] a variety of activities', required considerable planning to 'pitch a textbook (series) to pupils at the appropriate level' [S3]. This meant that authors had to understand 'the textbook as providing a basis for effective geographical learning' [P3] through engaging and clear explanations, the 'use of a variety of resources in interesting ways' [P2], and tasks which involved pupils in thinking geographically while checking that they understood the content and concepts involved.

Alongside this they also needed to understand the ways in which textbooks might be used in classrooms. One author noted that 'books, and activities, get used in very different ways by teachers. Some use the books heavily, others as a resource to pick up now and then' [S4]. As a result, this author's approach was to write and design the books 'to be self-contained/stand alone, so that a teacher can use them without access to other resources, and still deliver a good lesson' [S4]. This perspective identifies a challenge facing textbook authors, who may write with the intention that their textbook or series is used to provide a coherent course of study for pupils, even independently of the teacher, but who realise that it may be used in ways that cannot be fully accounted for in the structure of the book(s) or in the vocabulary, language and images of the text. It reinforces the point that the use of a textbook lies in the hands of the teacher or of departmental decision-making.

15.6.3 Developing Geographical Thinking Progressively

Authors noted clearly that in order to support pupils' geographical learning and thinking through a book and across a textbook series, they needed to structure and sequence their material to enable progression in learning. One secondary author made the point that planning for this was essential:

> In the initial planning phase for the series we spent a lot of time creating a KS3 development plan for the series…this plan worked towards pupil progression through KS3 with assessment opportunities clearly signposted and developed… [S2].

Other authors reinforced this approach, emphasising the need to understand the order through which pupils would work, how their knowledge and understanding would advance through sequential units in a book, and awareness of pupils' responses to and taking ownership at different phases in their schooling. Thus, constructing a textbook and series required that authors were clear about the level at which pupils will be working with the material:

> There is a huge difference in the way a Year 3 pupil would tackle the topic of water from the way in which a Year 6 pupil could deal with it [P2].

> Making sure to show a logical progression within a spread, and a chapter…. generally simpler material first, both within a chapter and across a series [S4].

Developing pupils' understanding of material was identified as a key point within the units of a book. To support the development of progressive learning, sequencing material was seen as a vital aspect of the process.

> The books are structured around units each of 3 double page spreads. The first spread introduces the topic with a strong image or arresting presentation. The second spread goes deeper and involves more research. The final spread focuses especially on fieldwork and

enquiry and tries to encourage teachers to undertake practical investigations with their pupils [P1].

Sequencing material does not in itself provide progression. Thus it is key that, as these authors stressed, 'a textbook series should be structured to develop progression in learning, and this can be done alongside demonstrating interconnections and coherence' [S2], having an underlying conceptual structure, and ensuring there is 'increased challenge within units and across the books' [S3]. This can represent a progressive 'journey' through the various aspects of curriculum geography. However, these authors recognised that 'progression is problematic, since teachers' use of the books in order, etc. cannot be guaranteed' [P1] for reasons noted above. Use of a textbook does not guarantee progression in pupils' subject learning.

15.7 Two Emergent Matters

This section picks up on two themes in this chapter. The first returns to the question of geographical knowledge and thinking and the connection with textbook writing in the 'eyes' of the authors in this sample. The second focuses on concerns about textbook use in classrooms and whether much is understood about their use in promoting geographical thought.

15.7.1 Textbooks and Geographical Thinking

Recent debates about the geography curriculum and teaching have explored the nature and meaning of its subject knowledge (Firth 2011, 2013; Morgan 2014; Young et al. 2014). The arguments have made limited direct reference to the role of geographical thought or thinking (Lambert and Morgan 2010; Roberts 2013, 2014). It seems assumed that subject thinking is implicit in the notion of subject knowledge. In effect, limited discussions of geographical thinking occur in parallel with those about knowledge (G.A. 2009; Morgan 2013). To an extent these draw on the perspectives of academic geographers about making sense of the world and of applying understanding in life (Jackson 2006; Bonnett 2008, Cloke and Johnston 2005; Henderson and Waterstone 2009; Nayak and Jeffrey 2011). Nonetheless, the arguments for subject knowledge and thinking are key to primary and secondary geography teaching, since they are fundamental to the purpose of pupils' school geographical learning: what is the point of a school subject if not to be better informed and to apply developing understanding? This is a perspective supported by this research. Though that use can be to achieve an extrinsic outcome, such as passing an examination, the real value for pupils lies in informing their lives and enabling them to live them more effectively though understanding what occurs in the world and its impacts, a rationale about which academic geographers and these textbook authors concur.

The Geographical Association's *Manifesto* argued that 'an essential educational outcome of learning geography is to be able to apply knowledge and conceptual understanding to new settings: that is to 'think geographically' about the changing world' (G.A. 2009, 9). This built on Jackson's argument that the purpose and value of geographical learning lies in applying the subject's concepts and ideas to enable 'us to see the connections between places and scales that others frequently miss' (Jackson 2006, 203). Geography offers a lens for understanding and thinking in our lives. The critical use of geographical concepts encourages awareness of the world around us and at a distance, providing an appreciation of how scale, for example, helps us to know the world in a variety of ways, such as seeing connections and commonality and diversity between places, environments and lives locally and internationally. It provides a range of perspectives to recognise the interrelationships and interdependences of places, lives and actions as they affect ourselves and others. This conceptual understanding is noted as geographical knowledge, distinguished from geographical information such as facts about particular resource extraction and uses or locating places on maps (G.A. 2009; Young et al. 2014). Its effect is thinking with geography. Again, this is a view affirmed by these textbook authors.

Morgan (2013, 280) suggests that such understanding supports our appreciation of and critical engagement with our everyday geography. It provides for 'multiple readings and meanings' to be recognised and debated about the complex issues and supposedly 'straightforward' solutions proffered in complex and controversial topics such as planning developments, access to resources, how places are valued, and environmental improvement and destruction. In exploring local and global concerns, these authors intended to engage pupils through their textbooks to grapple with such problems, recognising complexity and alternative solutions, and considering their means in their own lives and communities, making everyday connections. The application of concepts and engagement with differing perspectives and meanings is in the nature of geographical thinking and fosters personal understandings, stances, decision-making and, perhaps, actions.

For pupils to know and understand geography it is essential that they move beyond holding information about their world, and 'the world', and knowing about geography's concepts to applying them—to enable them to interpret, offer insights into and provide a 'different view'. This is necessary so that they 'ask questions about and investigate their own world' more widely and in greater depth and are able to critique their own and others thinking, ideas and 'solutions' (G.A. 2009, 11). In other words, developing geographical thinking is fundamental to enabling pupils' understanding of the subject, such that they recognise their own geographical nature, experience and perspectives to apply these in daily life more knowledgeably and thoughtfully. This was an intention engaged with by these textbook authors.

The purpose of school geography textbooks, appreciated by the authors in this study, is to enable the development of pupils' geographical knowledge and thinking not simply in the classroom context but usefully in daily life. To achieve this, they each set out to create a 'geographical world' (Freeman 2008) through which pupils could gain both information and insight by encountering and employing geographical concepts. Freeman argued that textbooks inevitably construct 'geographical worlds', influenced by a variety of factors, such as selected content, a book's place in a series, publishers' market interests, the particular experiences and world views of textbook writers, and the impact of external influences including curriculum requirements and examination syllabi. Some of the authors in this study viewed their textbooks as introducing pupils to particular aspects of the world, filtered, for instance, by the key stage 3 geography requirements. Others felt that it was significant that pupils encountered challenges and local and global issues and concerns, to understand that geographical knowledge is not simply a matter of information and that it deals with matters of contention. The authors' challenge to pupils was to encourage them to think these through. In this sense they were constructing 'geographical worlds' for pupils, chiefly as lenses for a deeper purpose, that pupils would develop understanding of the key geographical ideas which lay at the heart of the subject's thinking. An intention was to enable pupils to make such knowledge their own by taking up and applying a geographical lens.

The authors had a keen interest in promoting geography as a school subject and in doing this through engaging pupils with real, topical case studies which were stimulating and informative. This reflected authors' assumptions about pupils' lives and what might motivate their interests in the subject; they aimed to make connections to pupils' lives, to an extent countering Freeman's (2008) understanding that her sample of authors did not connect with pupils in this way. Yet this study's authors retained equally the view that textbooks were to inform pupils about geography, so as to help them understand its ways of perceiving and interpreting the world, such that they might apply this knowledge thoughtfully in understanding the environments of and events in their local area and the wider world.

In effect, this research has underscored that these geography textbook authors promote geographical thinking. It may be that this is an intention more widely held, but further research is needed to investigate the accuracy of such an interpretation. In this context it may also be useful to explore how they see the 'geographical worlds' they create through their texts as promoting geographical thinking to elucidate either a further sense of common ground or to recognise diversity in intentions and approaches. An international comparative study might investigate this most informatively.

15.7.2 Textbooks for Use

English authors write textbooks to be used in England's classrooms. This is the evident, if unmentioned, intention by the authors in this study. There has been concern about the use and impact of textbooks across subjects, and a variety of studies have investigated these concerns (Maughan et al. 2015), though recent interest has focused on aspects of reading, mathematics and science rather than on geography, which has received little attention (Newton and Newton 2008). Maughan et al.'s (2015) review and analysis draws equivocal conclusions and recommends that further, more systematic research is required. Indeed, it remains unclear how effective the use of textbooks by teachers and pupils is, whatever the intentions and qualities of the authors and textbooks. These reviews have been generated by government interest in fostering high quality pupil achievements nationally and in comparison to other nations (Gibb 2014; Oates 2014), but there has been little discussion about what this means, other than high scores in school examinations and international tests. As noted earlier, there is reference to improving subject information and knowledge but little if any comment appears about developing subject thinking and its application. Though there is brief reference to matters such as coherence between textbooks and curricula, notions of a 'good' textbook, design aspects of textbooks, and links to pedagogy and the classroom environment—as well as to issues of curriculum stability and textbook 'badging' to particular syllabi—there is a lack of examination of authors' perspectives on these matters.

In reviewing the qualities—and potential uses—of effective textbooks, Maughan et al. (2015, 23) suggest that such qualities include supporting teachers to develop thinking skills (for instance for the pupils of non-specialist primary teachers in geography), engaging with procedurally complex problems, using contexts and examples which draw pupils in and which they enjoy, ensuring clarity of their focus and what it is expected pupils should understand, illuminating the depth of expected standards, and providing evidently helpful instructional materials. Several of these aspects are supported by this study's geography textbook authors. They are concerned to ensure that pupils are clear about the geography they are learning and the expectations they are to meet. There is concern to provide a range of geographical contexts and problems from the straightforward to the complex, which require procedural thinking at different levels to utilise and enhance progress in pupils' geographical thinking. They stated that they wrote materials to achieve this, such as by providing stimulating case studies and examples to excite pupils about geography.

However, textbook authors, whatever their wishes, do not have influence directly over the context in which their textbooks are used, the atmosphere and nature of the classroom environment, nor the quality of teaching by whoever is using their textbooks. Neither does their publisher have any say in how many are bought and when they are used in schools. Publishers continue to produce geography textbooks and series for use in primary and secondary schools. There will be schools in which staff work through such books with pupils, as there will be schools where textbooks are sampled for particular geographical ideas and case studies (Lambert 1998, 2000). It may be that in some schools textbooks languish in cupboards, unused. The situation for geography textbooks remains unclear. That this study's authors believe in the use and value of textbooks is unquestioned, but it is not possible to state whether their intentions in writing textbooks are met, nor why this is the situation, without further research into geography textbook uses in primary and secondary geography lessons and courses.

What is evident is that the potential for textbooks as aids in learning remains strongly believed in, that the structure and sequencing of the ideas and content in geography textbooks is thought to be useful to help pupils' progression in understanding, and that they can enable pupils to develop their thinking geographically. To understand to what extent these beliefs are realistic and achieved is a matter for research into the uses of textbooks and, perhaps more importantly, into what effect they have on pupils' ideas about geography and the ways in which they may help develop and modify pupils' senses of geography—in what 'geographical worlds' they enable them to construct (Freeman 2008). While it is intimated that there is a culture of abjuring textbooks in classrooms (Truss 2013; Samuda 2005), it may be that this is less the case than it appears to be or is variable dependent on the quality of the textbooks and the capability of teachers in their planning, resource access and teaching (Lambert 2000). However, this is no simple matter since a variety of factors, including teaching quality and learner support, will be variables which impact on geographical learning. The uses and impacts of geography textbooks, therefore, need much deeper study to understand their role in enabling pupils to think geographically.

15.8 Conclusion

Those who write geography textbooks for primary and secondary teachers to use with their pupils do so for a variety of reasons (Waugh 2000; Freeman 2008). For primary authors motivation includes their perceived need to provide a good, accessible and informative resource about geography for non-specialist teachers to use. Secondary textbook authors write with geography specialists in mind, particularly for examination classes, though they are aware that some geography teachers of eleven to fourteen year olds may teach geography as an additional subject in their repertoire rather than as specialists. Thus secondary textbook writers may assume to a greater or lesser extent that teachers who use a textbook bring to it a moderate capability or strong expertise in geography, which authors for primary teachers do not. Nonetheless both sets of authors aim to fascinate and engage pupils to encourage and promote their interest in geography. They are concerned, therefore, to provide textbooks which are inviting, provide a rich variety of resources, contexts and case studies, with stimulating and challenging activities which make demands intellectually on pupils and which might equally open their minds to other perspectives, aspects of the world and issues than those which pupils may encounter in their daily lives (Cox 1973; Lidstone 1992). In part, they set out to achieve this by making connections appropriately with pupils' lives and experiences (Scott n.d.), while illustrating that such an approach informs teachers about how they might particularise these links for those they teach, perhaps encouraging fieldwork and other investigations through directed activities and independent study.

Though these authors worked in differing ways, some alone and others in teams, the fundamental characteristic of their rationales for writing geography textbooks was to develop pupils' knowledge and understanding of geography, as Fairgrieve (1937), Graves (2001) and Marsden (2001) have argued textbooks must do. Enabling pupils to think in informed and critically-aware ways about the world, across a range of its physical and human aspects and about places and nations, and to notice and seek to understand and explain spatial distributions and patterns, was the central curriculum focus and basis for teaching geography. It was clear from these authors' responses that their focus in geography textbook writing was in developing pupils' geographical thinking. As one author expressed it, 'the importance of the subject matter was a key influence' [P2], which another put more fulsomely as follows:

> Developing a framework of knowledge and understanding of the world, including place knowledge, knowledge and understanding of important processes and systems in the natural and human world….a range of geographical and other skills….(and) to consider and develop a range of values [S3].

Covering the requirements of national curriculum geography and of examination syllabuses accessibly for pupils and manageably for teachers, to engage, inform and challenge pupils, underpin this set of authors' approaches to writing geography textbooks. Authors felt that this required up-to-date knowledge of geography, the selection of stimulating and real places, environments and contexts to represent aspects of the world accurately, a variety of pedagogic approaches, challenging tasks and resources which provide a range of informative texts in themselves. Textbooks, they indicated, need to involve pupils in using the materials provided to enhance pupils' descriptive, analytic and evaluative thinking, set appropriately for the different age groups at whom the textbooks are aimed.

Authors are very conscious of the influences and constraints which support and constrain their writing. Their focus, they indicated, is steadfastly on encouraging pupils' geographical thought based in their knowledge and understanding of the world, such that they gained greater insight into life around them and further away, the natural processes and events of the earth and the interconnections and interdependence of people and the physical environment. For these textbook authors, engaging pupils and developing their geographical thinking was their central purpose. This perspective gives strength to geographical education. However, given the limited sample size of this study, authors' reasons for writing their geography textbooks needs to be further explored, as does the impact on pupils' knowledge and thinking of textbook use.

Acknowledgments This research was supported by a National Research Foundation of Korea Grant funded by the Korean Government (NRF-2013S1A2A1A01034085). It was conducted during Dr. Jongwon Lee's 2013–14 sabbatical at Oxford Brookes University, Oxford, United Kingdom, collaborating with Professor Simon Catling in the university's School of Education. The authors wish to acknowledge the contribution of the seven textbook authors who participated in the study.

References

Ahier, J. (1988). *Industry, children and the nation*. Lewes: Falmer Press.

Alexander, R. (2015). *Teaching to the text: England and Singapore*. http://cprtrust.org.uk/cprt-blog/teaching-to-the-text/. Accessed 21 March 2015.

Apple, M. (2014). *Official knowledge: Democratic education in a conservative age* (3rd ed.). New York: Routledge.

Bonnett, A. (2008). *What is Geography?*. London: Sage.

Bruillard, E. (2011). *Current textbook research in France: An overview*. Paper given at the International Textbook Symposium, April 28, Seoul, Korea: ITS.

Chambliss, M., & Calfee, R. (1998). *Textbooks for learning: Nurturing children's minds*. Oxford: Blackwell.

Cloke, P. & Johnston (Ed.) (2005). *Spaces of geographical Thought*. London: Sage.

Cox, B. (1973). Textbooks for secondary school geography. In D. Biddle & C. Deer (Eds.), *Readings in Geographical Education* (pp. 206–211). London: Whitcombe and Tombs.

D.E.S. (Department of Education and Science) (1991). *Geography in the national curriculum (England)*. London: Her Majesty's Stationary Office.

D.f.E. (Department for Education) (2013). *The national curriculum in England: Framework document*. www.gov.uk/dfe/nationalcurriculum. Accessed 25 Feb 2015.

D.f.E.E./Q.C.A. (Department for Education and Employment/Qualifications and Curriculum Authority) (1999). *The national curriculum for England: Geography*. London: Q.C.A.

Fairgrieve, J. (1937). *Geography in School* (4th ed.). London: University of London Press.

Fan, L. (2011). *Textbook research as scientific research: Towards a common ground for research on mathematics textbooks*. In Paper given at the International Conference on School Mathematics Textbooks, 12–14 Oct, Shanghai, China.

Firth, R. (2011). Debates about knowledge and the curriculum: Some implications for geographical education. In G. Butt (Ed.), *Geography, education and the future* (pp. 141–164). London: Continuum.

Firth, R. (2013). What constitues knowledge in geography? In D. Lambert and M. Jones (Eds.), *Debates in geography education*. (pp. 59–74). Abingdon: Routledge.

Firth, R. (2015). Constructing geographical knowledge. In G. Butt (Ed.), *MasterClass in geography education* (pp. 53–66). London: Bloomsbury.

Foster, S. (2011). Dominant traditions in international textbook research and revision. *Education Inquiry, 2*(1), 5–20.

Freeman, D. (2008). Creating geographical worlds: An investigation into the construction of school geography. *GeogEd, 2*(1). www.geography.org.uk/geoged. Accessed 3 Dec 2015.

Freeman, C., & Tranter, P. (2011). *Children in their urban environment: Changing worlds*. London: Earthscan.

Fuchs, E. (2011). Current trends in history and social studies textbook research. *Journal of International Cooperation in Education, 14*(2), 17–34.

G.A. (Geographical Association) (2009). *A different view*. Sheffield: The Geographical Association.

Gibb, N. (2014). *Quality Textbooks*. In Speech to The Publishers Association and The British Educational Suppliers Association Conference, 20 Nov, London, U.K. www.gov.uk/governmentspeeches/nick-gibb-speaks-to-education-publishers-about-quality-textbooks. Accessed 26 Feb 2015.

Gibb, N. (2015). *The social justice case for an academic curriculum*. Speech to The Policy Exchange, June 11, London, U.K. www.gov.uk/government/speeches/nick-gibb-the-social-justice-case-for-an-academic-curriculum. Accessed 12 June 2015.

Graves, N. (2001). *School textbook research: The case of geography 1800–2000*. London: Institute of Education.

Griffiths, J. (2013). *Kith: The riddle of the childscape*. London: Hamish Hamilton.

Hamann, B. (2007). Australia in German geography textbooks for middle schools. *International Research in Geographical and Environmental Education, 16*(2), 135–146.

Henderson, G., & Waterstone, M. (Eds.). (2009). *Geographic thought: A praxis perspective*. Abingdon: Routledge.

Holt, L. (Ed.). (2011). *Geographies of children, youth and families: An international perspective*. Abingdon: Routledge.

Hopkin, J. (2001). The world according to geography textbooks: Interpretations of the English national curriculum. *International Research in Geographical and Environmental Education, 10*(1), 46–67.

Jackson, P. (2006). Thinking geographically. *Geography 91*(3), 199–2014.

Janko, T. & Knecht, P. (2013). Visuals in geography textbooks: Categorization of types and assessment of their instructional qualities. *Review of International Geographical Education Online, 3*(1), www.rigeo.org. Accessed 14 June 2013.

Lambert, D. (1998). Rediscovering textbook pedagogy: The need for research on the use of textbooks in geography classrooms. In A. Kent (Ed.), *Issues for research in geographical education: Research forum 1: Textbooks* (pp. 23–38). London: International Geographical Union Commission on Geographical Education/Institute of Education, London University.

Lambert, D. (2000). Textbook pedagogy—Issues on the use of textbooks in geography classroom. In C. Fisher & T. Binns (Eds.), *Issues in geography* (pp. 108–119). London: Roudledge.

Lambert, D., & Morgan, J. (2010). *Teaching geography 11–18*. Maidenhead: Open University Press.

Lee, J., & Catling, S. (2016). Some perceptions of English geography textbook authors on writing textbooks. *International Research in Geographical and Environmental Education, 25*(1), 50–67.

Lidstone, J. (1992). In Defence of Textbooks. In M. Naish (Ed.), *Geography and education: National and international perspectives* (pp. 177–193). London: Institute of Education, University of London.

Long, M., & Roberson, B. (1966). *Teaching geography*. London: Heinemann Educational Books.

Marsden, W. (1988). Continuity and change in geography textbooks: Perspectives from the 1930s to the 1960s. *Geography, 73*(4), 327–343.

Marsden, W. (2001). *The school textbook: Geography, history and social studies*. London: Woburn Press.

Martin, F. (2006). *Teaching geography in primary schools*. Cambridge: Chris Kington Publishing.

Maughan, S., Smith, J. & Hamer, J. (2015). *A Literature review of textbook use and links to educational standards: Final report*. Manchester: Alpha*Plus* Consultancy Ltd. www.publishers.org.uk/_resources/assets/attachment/full/0/19818.pdf. Accessed 20 Nov 2015.

Morgan, J. (2003). Imagined country: National environmental ideologies in school geography textbooks. *Antipode, 35*(4), 444–462.

Morgan, J. (2012). *Teaching secondary geography as if the planet matters*. Abingdon: Routledge.

Morgan, J. (2013). What does it mean to think geographically? In D. Lambert & M. Jones (Eds.), *Debates in geography education* (pp. 273–281). Abingdon: Routledge.

Morgan, J. (2014). 'Neither existance nor future': The social realist challenge to school geography. In B. Barrett and E. Rata (Eds.), *Knowledge and the Future of the Curriculum: International studies in social realism*. (pp. 136–152). Basingstoke: Palgrave Macmillan.

Nayak, A., & Jeffery, A. (2011). *Geographical thought: An introduction to ideas in human geography*. Harlow: Pearson.

Newby, P. (2014). *Research methods for education* (2nd ed.). Harlow: Pearson.

Newton, L., & Newton, D. (2008). To what extent can children's geography books help a primary school teacher explain cause and effect? *International Research in Geographical and Environmental Education, 15*(1), 29–40.

Norcia, M. (2010). *X marks the spot: Women writers map the empire for British children, 1790–1895*. Athens: Ohio University Press.

Oates, T. (2014). *Why textbooks count: A policy paper*. In Paper given at The Publishers Association and The British Educational Suppliers Association Conference, London,

November 20. Cambridge Assessment. www.cambridgeassessment.org.uk/Images/181744-why-textbooks-count-tim-oates.pdf. Accessed 23 Mar 2015.
Peled-Elhanan, N. (2012). *Palestine in Israeli school books: Ideology and propaganda in education*. London: I.B. Taurus.
Q.C.A. (2007). *The National Curriculum 2007*. London: Q.C.A. www.teachfind.com/qcda/national-curriculum-statutory-requirements-key-stages-3-and-4-september-2008-pdf. See also: *Geography for key stage 3 programme of study and attainment target*. www.rgs.org/NR/rdonlyres/2FFCE97B-5753-4A98-9291-6E652E3185AD/0/FW_GeographyNC.pdf.
Rawding, C. (2013). *Effective innovation in the secondary geography curriculum*. Abingdon: Routledge.
Roberts, M. (2013). *Geography through enquiry: Approaches to teaching and learning in the secondary school*. Sheffield: Geographical Association.
Roberts, M. (2014). Powerful knowledge and geographical education. *The Curriculum Journal, 25* (2), 187–209.
Robson, C. (2011). *Real world research* (3rd ed.). Chichester: Wiley.
Samuda, V. (2005). Expertise in pedagogic task design. In K. Johnson (Ed.), *Expertise in second language learning and teaching* (pp. 230–253). New York, NY: Palgrave MacMillan.
Scoffham, S. (Ed.). (2010). *Primary geography handbook*. Sheffield: Geographical Association.
Scott, J. (n.d.). *The teaching of geography in Canadian elementary schools*. Alberta: The Collegiate Press.
Shulman, L. (2004). *The wisdom of practice*. San Francisco: Jossey-Bass.
Somerville, M., & Green, M. (2015). *Children, place and sustainability*. Basingstoke: Palgrave Macmillan.
Spencer, C., & Blades, M. (Eds.). (2006). *Children and their environments: Learning, using and designing spaces*. Cambridge: Cambridge University Press.
Truss, E. (2013). *In defence of the textbook*. In A speech delivered at the Education Publishers' Conference, Institute of Physics, London, on December 2nd. www.gov.uk/government/speeches/elizabeth-truss-speaks-to-education-publihsers-about-curriculum-reform. Accessed 20 Dec 2013.
Walford, R. (1995). Geographical textbooks 1930–1990: The strange case of the disappearing text. *Paradigm, 18*. http://faculty.education.illinois.edu/westbury/Paradigm/Walford.html. Accessed 26 Jan 2015.
Waugh, D. (2000). Writing geography textbooks. In C. Fisher & T. Binns (Eds.), *Issues in geography teaching* (pp. 93–107). London: Routledge.
Westaway, J., & Rawling, E. (1998). An analysis of key stage 3 geography textbooks. *Teaching Geography, 22*(1), 36–38.
Wright, D. (1996). Textbook research in geographical and environmental education. In M. Williams (Ed.), *Understanding geographical and environmental education: The role of research* (pp. 172–182). London: Cassell.
Young, E. (1977). A walking-stick, not a crutch. *Teaching Geography, 2*(4), 173–175.
Young, M. (2008). *Bringing knowledge back in*. Abingdon: Routledge.
Young, M., Lambert, D. with Roberts, C. & Roberts, M. (2014). *Knowledge and the future school*. London: Bloomsbury.
Young, M., & Muller, J. (2016). *Curriculum and the specialisation of knowledge: Studies in the sociology of education*. Abingdon: Routledge.

Part IV
Conclusion

Chapter 16
Reflecting on What Makes Geographical Thinking Powerful

Clare Brooks, Graham Butt and Mary Fargher

In his seminal article *Thinking Geographically*, Jackson (2006) concludes with the following:

> Thinking geographically is a uniquely powerful way of seeing the world. While it does not provide a blueprint … thinking geographically does provide a language – a set of concepts and ideas – that can help us see the connections between places and scales that others frequently miss. That is why we should focus on geography's grammar as well as on its endless vocabulary. That is the power of thinking geographically *(p9).*

Jackson's view, that geographical thinking is inimitably significant for education has also been the central thread running throughout *The Power of Geographical Thinking*. The purpose of our collective response has been to argue that geography as a body of knowledge offers young people an extraordinary and vital perspective with which to view and understand their world. Not a surprise one may argue considering that all of the authors are geographers, geography educators and geography teachers. However there are also a number of important, broader themes that can be identified from the chapters included here which are useful to articulate in these final pages. Specifically they include ideas concerning powerful thinking and knowledge, powerful thinking and pedagogy and the wider implications of the power of geographical thinking and its educational role.

First, with regards to ideas concerning powerful thinking and powerful knowledge, there is no doubt that the chapters in this volume represent a strong collective call from the represented authors for 'bringing knowledge back in' (Young 2008) to geography education. A convincing theoretical argument has been advanced for

C. Brooks (✉) · M. Fargher
UCL, Institute of Education, London, UK
e-mail: c.brooks@ucl.ac.uk

M. Fargher
e-mail: m.fargher@ucl.ac.uk

G. Butt
Oxford Brookes University, Oxford, UK
e-mail: gbutt@brookes.ac.uk

C. Brooks et al. (eds.), *The Power of Geographical Thinking*,
International Perspectives on Geographical Education,
DOI 10.1007/978-3-319-49986-4_16

geographical thinking being a form of powerful knowledge (Butt, Chap. 1), with the reader also provided with critical 'food for thought' on what kind of analytical frameworks geography teacher educators and teachers may use in applying this thesis in the geography classroom (Maude, Chap. 2; Uhlenwinkel, Chap. 3). The editors would argue that whilst these conceptual debates are on-going, this book has discussed in new ways how the power of geographical thinking for young people can become an integral part of their education in schools (Roberts 2011).

This is not to say that the authors here have spoken with one advocatory voice about the *educational value* of thinking geographically. As one would expect, their varied interpretations reflect the perspectives of the wider geography education community on the issue. For example, with regards to developing geographical thinking linked to concepts of sustainability, some are more strongly supportive of developing young people's role *through geography* as global citizens. Others strongly believe that moving geography education in this direction weakens it, taking learners away from the discipline itself. In a similar way some authors in this volume have identified the value of geography as a 'bridging subject', which can be used in an inter-disciplinary way to extend the power of geographical thinking into other subjects in school-based education.

This book has provided a collective rationale which contributes to the on-going debates concerning the development of geographical thinking as a cornerstone in school geography, but also in education per se. Authors have agreed that powerful geographical thinking should include geography's core concepts (Lambert 2011) but that core knowledge is not enough alone in developing young peoples' geographical thinking. Geography is much more than learning facts and significant skills (Butt 2011). Whilst geography has no doubt benefited from the 'knowledge turn' (Young 2008), thinking geographically goes beyond that and requires a conceptual understanding of the academic discipline of geography.

Second, with regards to the ways in which powerful geographical thinking can be applied through pedagogy, this volume provides new and significant theory and evidence-based knowledge for geography educators and teachers. It achieves this goal in a number of ways. For example: through modelling how teachers can use powerful geography knowledge concepts such as space and place through their pedagogy (Fogele, Chap. 4); through connecting teacher's geographical knowledge directly with their pupils' powerful geographical thinking (Arenas-Martija et al., Chap. 5); through critical consideration of the potential and limitation of using a technology such as GIS to construct geographical knowledge (Fargher, Chap. 10) and the importance of writing textbooks which focus on the development of powerful geographical thinking (Catling and Lee, Chap. 14).

Third, there is a strong message articulated with regards to the role of powerful geographical thinking and its implications for wider society through the value of education itself (Brooks, Chap. 11). The choice of examples used by authors strongly reflects the influence of a rapidly changing world and geography education's role in helping young people to make sense of it. Importantly in a volume which spans contributions from across the globe, the reader has also been given the chance to reflect on the role that local context plays on shaping geographical thinking (Salinas et al., Chap. 12). Other chapters have focussed on several significant contemporary themes which have wider implications for education and society as a whole. These have included climate change (Chang and Pasqua, Chap. 6), sustainability (Palacios et al., Chap. 7) migration (Martins, Chap. 13), the critical use of technologies in education (Fargher, Chap. 10) and society as a whole. These choices of foci make timely and important contributions to the wider debate on significance of geographical thinking and an education for the future (Butt 2011).

Key education stakeholders are complicit and responsible for the kind of geographies that are created in schools. Massey and Clark make a similar claim:

> Whether we like it or not, our actions (and our inactions) have effects. Sometimes in big ways, more often in small ways, we are implicated in the production of this world (Massey and Clark 2008, p. 1).

In a book such as this, which encompasses a broad remit with respect to geographical thinking, it's conceptual basis, pedagogies for geographical thinking and the wider societal implications of developing young people's powerful geographical thinking, it is not surprising that some elements of our aims remain unfulfilled. However there is a consensus that concurs with Young's thesis that powerful disciplinary knowledge in schools (both within geography and elsewhere) must take pupils beyond the everyday. It is arguably not for the authors of this book to say that the everyday cannot be of value in thinking geographically, but many would make the case that the everyday is not enough to make geographical thinking powerful.

In almost every chapter of this volume, we can see that powerful geographical knowledge is hard won. It requires teachers to (re)engage with the very roots of geography's major conceptual contributions to knowledge about our world. This requires what is also identified as a common theme throughout the book—a need for geography teachers who have been educated to be knowledgeable, to be critical, to be thoughtful, and to be skilful in developing pedagogies which hone the development of powerful geographical thinking in our young people.

> If that exceedingly heterogenous group of people called 'geographers' have anything in common it is this (and it is inevitably generic, even banal): they are together engaged in an on-going process of producing, sharing, reconstituting and distributing knowledge. This does not make geography a purely epistemological enterprise; on the contrary, the geographical knowledges that are our stock-in trade both arise from and inform our practical engagements with the world (Castree et al. 2008, p. 680).

References

Butt, G. (ed.) (2011). *Geography, education and the future*. London: Continuum.
Castree, N., Fuller, D., Kent, A., Kobayashi, A., Merrett, C. D., Pulido, L. & Barraclough, L. (2008). Geography,pedagogy and politics. *Progress in Human Geography, 32*(5), 680–718.
Jackson, P. (2006). Thinking geographically. *Geography, 91*(3), 199.
Lambert, D. (2011). Reviewing the case for geography, and the 'knowledge turn' in the English National Curriculum. *Curriculum Journal, 22*(2), 243–264.
Massey, D., & Clark, N. (2008). Introduction. In N. Clark & D. Massey & P. Sarre (eds.) *Material geographies: A world in the making*. London: Sage.
Roberts, M. (2011). *What makes a good geography lesson?* Geographical Association. Available at www.geography.org.uk/download/ga_prmghwhatmakesageographylessongood.pdf1. Accessed 03 June 2016
Young, M. (2008). From constructivism to realism in the sociology of the curriculum. *Review of research in education, 32*(1), 1–28.

Printed by Printforce, the Netherlands